Human Factors and Ergonomics of Prehospital Emergency Care

Human Factors and Ergonomics of Prehospital Emergency Care

Edited by
Joseph R. Keebler
Elizabeth H. Lazzara
Paul Misasi

CRC Press
Taylor & Francis Group
Boca Raton London New York

CRC Press is an imprint of the
Taylor & Francis Group, an **informa** business

CRC Press
Taylor & Francis Group
6000 Broken Sound Parkway NW, Suite 300
Boca Raton, FL 33487-2742

First issued in paperback 2019

ISBN-13: 978-1-4822-4251-5 (hbk)
ISBN-13: 978-0-367-87011-9 (pbk)

Library of Congress Cataloging-in-Publication Data

Names: Keebler, Joseph R. | Lazzara, Elizabeth. | Misasi, Paul.
Title: Human factors and ergonomics of prehospital emergency care / Joseph R. Keebler, Elizabeth H. Lazzara, Paul Misasi.
Description: Boca Raton, FL : Routledge/CRC Press, 2017. | Includes index.
Identifiers: LCCN 2016026220| ISBN 9781482242515 (hardback) | ISBN 9781482242522 (ebook)
Subjects: LCSH: Emergency medical services--United States | Emergency medicine--United States. | Medical emergencies--United States. | Emergency medical technicians--United States. | Medical errors--United States--Prevention. | Patients--Safety measures.
Classification: LCC RA645.5 .K44 2017 | DDC 362.18--dc23
LC record available at https://lccn.loc.gov/2016026220

Visit the Taylor & Francis Web site at
http://www.taylorandfrancis.com

and the CRC Press Web site at
http://www.crcpress.com

Contents

Foreword

It was March 28, 2011, when the director of Sedgwick County Emergency Medical Service (EMS) called me into his office. I had been expecting his call, but was none-theless quite nervous. I had applied for a promotion—a big one, one that would take me from being a field medic and turn me into "brass" (an administrator). At the time I was a lieutenant of six years and had been on the street for a total of eight. I had applied for promotions before, five to be exact, and I had been turned down every time. Prior to this opportunity, I wanted what I think most field medics want at some point in time, which is to be a field supervisor, to be in charge, and drive around in a Suburban instead of an ambulance, to be a supermedic—make the cool calls, do the cool stuff, have extra privilege and authority, be "Captain Misasi." Sedgwick County EMS was in a period of substantial attrition. Many of the field supervisors who had been at their post for almost two decades were beginning to retire. There was finally an opportunity for upward movement in the organization. Over the course of a few years, I had applied, tested, and retested for the rank of captain and even made my way to the last cut before the interviews once, but was not invited to do so. Befuddled and angry each time, I wondered why I had been passed over. Seeking input from my supervisors and peers did not seem to be much help, as no one seemed to be able to identify any major character flaws and it certainly was not for lack of education. I had obtained a master's degree in Emergency Health Services from the University of Maryland, Baltimore County, in 2008. The common theme of reasons why I did not make the cut was simply that I did not have enough time in the saddle. I was, after all, only 28 years old. Perhaps I was a bit naive.

It took me four years to get through graduate school on a part-time basis while working as a street medic, but education and persistence are in my blood. My mother immigrated to the United States from Mexico when she was four years old, along with her eight brothers and sisters. Her father had come to the United States a couple of years earlier on a work visa and was employed by a co-op in Kiowa, Kansas. They had nothing when they moved to the States, they lived on dirt floors, and to this day my mother will eat nearly every part of a chicken, including the feet, and will suck the marrow out of any bones. I have never been that hungry, but she had been many times. My mother was a successful family medicine physician in Wichita, Kansas, with many loyal and adoring patients. My brother and I grew up in the nursing sta-tions of hospital wards and offices of her clinic, running amok, making paperclip chains, and photocopying about anything we could get our hands on. Her achieve-ments in life truly are the manifest of the American Dream. Her father worked two jobs to support the family, and whenever she asked for permission to get a job, her father's response was always "your job is to get an education." I heard those words many times in my life.

As I progressed through my education in emergency health services, taught by the greats of the EMS world, Stephen Dean, Rick Bissell, Brian Maguire, Skip Kirkwood, and Mike Taigman, to name a few, I began to observe and consider the gap between what a great EMS agency could be and what mine was. Hearing story after story of how coworkers were punished for making mistakes and even making a few of my own, my frustration grew, which I channeled into productive motivation to keep me going through all the long hours of studying and working. I was going to be better than "them" at "building one" (administrative headquarters for the service) because I knew in my gut there had to be a better way to manage a service and its people. Time after time we would all receive paperwork for doing (or not doing) things that felt as if we had been set up to do, to fail, because sometimes it was just damned impossible to do things like they were "supposed" to be done. Do not get me wrong; I loved what I did as a street medic. I most loved the satisfaction of doing my job well *despite* all the obstacles I faced. Paramedics are, after all, the masters of innovation and adaptation. However, there is nothing more frustrating than a computer that will not do what it is supposed to; a pulse oximeter alarm that sounds every time a patient moves their finger; an electrocardiogram tracing that is so full of artifact that you decide as long as you can make out some R waves, the patient is doing okay; an air conditioner that blows hot air; or a radio that waits to give you a low battery chirp right as you are trying to call in the alert for your level I trauma patient. Still, I wanted to move up in my organization to prove just that, that there really was a different way of managing, of leading and supporting paramedics on the front lines. I made it my personal mission to do so.

So there I sat, across from the service director in his office, nervously awaiting news of whether I had been picked among one other candidate for the vacant Deployment Manager position (which held a captain rank). I vaguely remember Scott recapping my performance in the interview and what he was looking for in someone filling this role, but I distinctly remember him offering me the job, to which I replied, "I was afraid you might say that." It was a life-changing moment.

The following Monday was my first day in the office. After my first staff meeting, I remember returning to my office and putting my head on my desk. I asked myself several times, "what the hell have you done?" and nearly panicked, desperate for the safety and comfort behind the windshield of an ambulance. A few months later, the deputy director called me and the other division officers into a discussion about what appeared to be a problem with medication errors, specifically dosing problems that were being reported with fentanyl. The text of the meeting was as follows:

Colleagues,

I seek a fresh perspective and some clarity regarding recent patient and provider safety events that are occurring in our areas. Each of you, and I myself have voiced uneasiness and are frankly puzzled about how some of these circumstances arise. Over this holiday weekend, give some intentional consideration about what are the factors and variables that influence the development of untoward events. It is time we transition through the documentation and discipline phases and move into root cause analysis and course correction.

I will assume the following; our employees are educated and committed to caring for their patients, there is not a single factor, this likely has roots in all divisions hence all are represented and asked to assist, this dialogue is a safe place to ask the probing questions necessary, that we seek an even better way to promote and ensure the safety of our patients and providers, and that we are a learning institution and as such, consider this an exciting challenge and chance to enjoy new understandings.

As we all are busy and needing help in prioritization of initiatives, I intend to bring this forth next Tuesday at the morning Staff Meeting and will identify July 29, 2011 as the date on which we will have completed our investigation, analysis, and design of a course of correction for approval and implementation.

It was the phrases "I seek a fresh perspective" and "it is time we transition through the documentation and discipline phases and move into root cause analysis and course correction" that captured my attention, having been on the receiving end of formal discipline for errors I had never intended to commit. And, I certainly did not want to be a part of an administration that continued the same reactionary, punitive cycle that had never produced the desired results. As Quint Studer states in his book *Results That Last*, "no organization has ever reprimanded its way to success." So this was it; I was going to solve the problem of medication errors and prove that there was a better way.

Five days later, the Deputy Director e-mailed me and the other division officers a hyperlink to a PDF copy of the book *The Field Guide to Human Error Investigations* by Sidney Dekker.* It was exactly what I needed to begin understanding how to even approach a problem such as a medication error. What I did not know was that the words of this text would change my life. The author's words were as if someone had articulated the rationale behind what I had felt intuitively as a street medic frustrated by story after story of paramedics being reprimanded. I recall many times as I read the words of this book that I wanted to stand on my desk and shout "Yes!" for Dr. Dekker had opened my eyes to an entire field of science that I nor many other people in the world even know exists—the science of human factors.

As an undergraduate at Oklahoma State University (OSU) in the aviation education program, I had been introduced to human factors without even knowing it, in fact, I had an entire semester course in human factors without even knowing it—the course was called Cockpit Resource Management (now known as Crew Resource Management). I entered my first semester of college at OSU as an 18-year-old private pilot, and as most pilots can attest, it takes only one flight at the helm of an aircraft and you are an addict. My plan was to graduate with a bachelor's degree in aviation education and work as a corporate or airline pilot. I earned an instrument rating, complex endorsement, and had taken and passed the commercial pilot written exam when life steered me toward EMS, leaving aviation behind; or so I thought.

As I read Dekker's book, I had vague recollections of my cockpit resource management education and I began to wonder if somehow, the concepts were transplantable to EMS. It became clear that we paramedics operated in nearly the same type of high-pressure technical work environments and faced the similar challenges and risks that required us to be prepared at any time for emergency action and we were rarely alone as field providers; that is, we always have a copilot. Even though paramedics and pilots work in teams of dyads most of the time, pilots and paramedics are trained many places still today with the same mind-set: "someday you will be the only one who knows what to do so you have to be able to handle everything on your

---

* My apologies to Dr. Dekker and the publishers of this text. I am not sure whether the version I was given and read was a pirated copy or not, but I assure you, had I not read it, I would not have bought another eight or so of his books.

own and you need to be able to tell everyone else what to do." Slowly, everything began to click and this human factors stuff seemed to go together with the work of EMS like chocolate with peanut butter.

My colleague Dave, the newly promoted operations manager, and I attended the Mile High Regional Emergency Medical and Trauma Advisory Council Safety Summit in Loveland, Colorado, in July of 2011, where we heard presentations from Scott Griffiths, the former head of safety for American Airlines and who was at the time working for a company called Outcome Enginuity, which was founded by David Marx, an engineer turned lawyer and author of the book *Whack-a-Mole: The Price We Pay for Expecting Perfection*. (David Marx also happens to have worked with Dr. James Reason, an influential cognitive psychologist who has produced much literature on the nature of human error and theories of accident causation.) Griffiths's talk was incredibly compelling, and he was followed by Paul LeSage, another excellent speaker and (co)author of a book that had recently been lent to me, *Crew Resource Management: Principles and Practice*. In fact, I had the book there with me, read it cover to cover during that time, and asked LeSage to sign it ... to me (which meant I had to buy another copy to replace it for the person who had lent it to me in the first place). The stars were beginning to align for me and my work as a budding middle manager in a large EMS system (by Kansas standards). I read everything I could get my hands on about the subject of human factors. I began searching for educational opportunities on the subject and found that one of the nation's top schools for human factors psychology was right in my backyard, Wichita State University. I made an appointment to visit the department head, and the rest, as they say, is history. As I write this text today, I am ending my second year as a doctoral graduate student and will likely be beginning my third at the time of this book's release.

I am now the Clinical Division Manager for Sedgwick County EMS, and I can tell you that it is nearly impossible to find an aspect of the delivery of EMS that cannot be improved or affected positively when we perfuse our work with the science of human factors. Whether we know it or not, it permeates nearly everything we do, from the design of clinical decision support to the design of a patient care reporting interface, to investigating the failure of a crew to recognize subcutaneous emphysema and decompress a tension pneumothorax, to understanding a paramedic's decision to prolong scene time on a critical trauma to intubate the patient, or simply the visual representation of quantitative information. Learning about this field lays forth a path and provides the tools to achieve improvements in aspects of clinical care, logistics and operations, patient and provider safety, and much, much more. Anywhere a human mind must interact with information in the environment to achieve some goals, human factors are at work.

Thankfully, in my position I have been given a substantial amount of autonomy to direct my own work, which has enabled me to demonstrate the benefits of the human factors approach, having made substantial contributions to patient safety through improvements in cot (stretcher) operations preventing patient drops; a reduction in medication errors by 49% (71% with fentanyl alone) over a 54-month period; refining the investigation and analysis of accidents and incidents; and helping transform the culture of our organization into a high-reliability, metalearning organization.

Having consumed a large volume of human factors literature in a short amount of time, I realized the importance and the impact that it could have not just in EMS, but in medicine as well. Although the domain of medicine was introduced to human factors and the need for systems engineering in the 1990s, it has only slowly begun adopting and appreciating the benefits it affords, and in some cases, way too late (just ask your local emergency room doctor what they think of the usability of their electronic medical records). And, unfortunately, the situation that EMS providers are all too familiar with, EMS has (until now) all but been forgotten. So as human factors science works to convince the medical community of its value and importance in healthcare, I decided I would do whatever I could to get the attention of human factors and direct it to EMS; this book is the result. EMS is ready for human factors, but we first have to understand just what it is all about and how we can leverage it to improve our systems of delivery of prehospital medicine, from complexity and resilience to that pesky problem of human error.

Thus, the purpose of this book (although far from comprehensive) is to serve as the front door to a world of scientific investigation and open the reader's mind to the myriad of possibilities and novel solutions crafted upon the tenets of human factors that can change EMS for the better. As Dr. Yuval Bitan writes in Chapter 12, EMS is uniquely poised and ready to show the broader medical domain the things they should have learned a decade ago and improve upon what aviation learned four decades ago.

It is our hope that by reading this book, field providers will be inspired to reconsider the problems they face every day from a new perspective and evaluate alternative ways of doing their important work safely and effectively, that leaders in EMS will ask new questions about the problems they face in designing and managing systems and people, and that researchers will turn their attention to a domain of work that is desperate for their investigation and intervention.

Paul Misasi, MS, MA, NRP
Clinical Division Manager, Sedgwick County EMS

Editors

Joseph R. Keebler, PhD, has over ten years of experience in conducting experimental and applied research in human factors, with a specific focus on training and teamwork in military, medical, and consumer domains. Dr. Keebler currently serves as an assistant professor of human factors and systems at Embry-Riddle Aeronautical University, Daytona Beach, Florida. He has led projects aimed at the implementation of human factors in complex, high-risk systems to increase safety and human performance. This work includes command and control of teleoperated unmanned vehicles, communication and teamwork in medical systems, and development of simulation and gamification of training for advanced skills, including playing the guitar and identifying combat vehicles. Dr. Keebler's work includes over 50 publications and over 60 presentations at national and international conferences.

Elizabeth H. Lazzara, PhD, prior to being an assistant professor at Embry-Riddle Aeronautical University, Daytona Beach, Florida, earned her doctorate in applied experimental human factors psychology from the University of Central Florida, where she also conducted research at the Institute for Simulation and Training in Orlando, Florida. Although she has had extensive experiences in military, academic, and commercial settings, Dr. Lazzara's primary interests lie in improving the quality of patient care within the healthcare domain. She strives to make a long-lasting and significant impact by examining and advancing the science and practice of clinical care and patient safety issues pertaining to human performance, teamwork, team training and simulation-based training, and performance measurement.

Paul Misasi, MS, MA, has been a paramedic for 13 years and currently serves as the Clinical Manager of Sedgwick County Emergency Medical Service, Wichita, Kansas. He holds a master of science degree in emergency health services from the University of Maryland, Baltimore County, a Master of Arts in Psychology and a bachelor of science degree in health service organization and policy, both from Wichita State University. He is the first paramedic/ambulance service manager to achieve board certification in patient safety through the National Patient Safety Foundation. Prior to becoming a paramedic, Misasi began his study in human factors through his flight training and pilot licensure at Oklahoma State University and is now a third-year doctoral student in human factors psychology at Wichita State University. He is principal developer of the industry-wide best practice of the Medication Administration Cross-Check© protocol, which is under peer review as the first empirically validated medication verification process.

Contributors

Wendy L. Bedwell, PhD, joined the faculty of the University of South Florida, Tampa, Florida, through the Industrial–Organizational Psychology program as an assistant professor upon receiving her PhD from the University of Central Florida. Her research is focused on understanding the factors that influence learning and adaptation in individuals and interdisciplinary and/or multicultural teams operating in complex, dynamic, and extreme environments. She also considers training effectiveness, with the goal of tying training techniques, tools, and components to desired learning outcomes. Dr. Bedwell conducts both laboratory- and field-based research and has worked with medical, military, and other professional populations. She currently has grants with the National Aeronautics and Space Administration that are focused on understanding compositional and team process factors that influence effectiveness over time. Prior to receiving her PhD, Dr. Bedwell worked as a trainer and instructional designer in academic, business, and government settings. These experiences provide the foundation for her scientist-practitioner approach to research and doctoral student education.

Lauren E. Benishek, PhD, is an industrial/organizational psychologist working as a research fellow at the Johns Hopkins University School of Medicine's Armstrong Institute for Patient Safety and Quality, Baltimore, Maryland. Dr. Benishek's specialties and research interests coalesce on three major themes: (1) professional talent development, (2) safety and well-being, and (3) teamwork and unit dynamics. Her work includes efforts to improve medical care quality and safety via specific projects involving national and international collaborations to measure healthcare safety culture and clinical outcomes, develop and evaluate training programs, improve unit dynamics, and enhance employee well being. She has coauthored 12 peer-reviewed journal articles, four book chapters, one book, and 26 conference presentations. In addition, Dr. Benishek has chaired four symposia and delivered 17 invited talks. Her work appears in various outlets including *The Joint Commission Journal on Quality and Patient Safety, Simulation in Healthcare,* and *Medical Science Educator.*

Yuval Bitan, PhD, is a researcher and a human factors engineer, specializing in ways to improve patient safety through the implementation of human factors principles in the healthcare complex working environment. Dr. Bitan was a research associate at the Cognitive Technologies Laboratory (University of Chicago, Illinois), HumanEra (University Health Network, Toronto, Canada), and the University of Toronto. Dr. Bitan holds a PhD in human factors engineering from the Ben-Gurion University of the Negev, Beer-Sheva, Israel, where he teaches at the Department of Industrial Engineering and Management and leads the Human Factors in Healthcare laboratory.

Ian Coffman, PhD, is a research scientist at AnthroTronix, Silver Spring, Maryland. Dr. Coffman received his doctorate in cognitive science from The Johns Hopkins University. He specializes in both theoretical and applied cognitive testing and language sciences (such as phonetics speech perception, and theoretical and experimental phonology). Dr. Coffman's human factors background centers on task analysis, alarms and warnings, and natural language interfaces. He also has extensive experience with regression modeling, including hierarchical models and Bayesian methods.

Julius Cuong-Pham, MD, PhD, is an associate professor of emergency medicine and anesthesia critical care at the Johns Hopkins University School of Medicine, Baltimore, Maryland. His recent research involves studying adverse events, medication errors, adverse event reporting systems, medical device safety, and healthcare-acquired infections. Dr. Cuong-Pham's research has been funded by the Agency for Healthcare Research and Quality, the Robert Wood Johnson Foundation, the National Patient Safety Agency, the Canadian Patient Safety Institute, the National Library of Medicine, the World Health Organization, and other private organizations. Dr. Cuong-Pham provides clinical care in both in emergency department and the intensive care unit. This cross-training in two clinical disciplines provides a unique perspective on the presentation and progression of patient illness. In addition to his clinical duties, Dr. Cuong-Pham provides medical education to fellows, residents, and medical students.

Sidney Dekker, PhD, graduated from Ohio State University in 1996, is professor of humanities and social science at Griffith University in Brisbane, Australia, where he runs the Safety Science Innovation Lab. He is also a professor (honorary) of psychology at the University of Queensland, and a professor (honorary) of human factors and patient safety at the Royal Children's Hospital in Brisbane. Previously, he was a professor of human factors and system safety at Lund University in Sweden. After becoming a full professor, he learned to fly the Boeing 737, working part time as an airline pilot out of Copenhagen. He has won worldwide acclaim for his groundbreaking work in human factors and safety and is the best-selling author of, most recently, *The Field Guide to Understanding "Human Error"* (2014), *Second Victim* (2013), *Just Culture* (2012), *Drift into Failure* (2011), and *Patient Safety* (2011). His latest book is *Safety Differently* (2015).

Harinder Dhindsa, MD, MPH, MBA, is an emergency medicine physician at Virginia Commonwealth University Health in Richmond, Virginia. He is an associate professor and the chief of the Department of Emergency Medicine. He also serves as the Medical Director for Virginia Commonwealth University Health's critical care transport program. He received his medical degree from the University of Maryland School of Medicine and has been in practice for more than 25 years. Dr. Dhindsa completed his residency in emergency medicine at George Washington University and Georgetown University. He completed an emergency medical services fellowship at George Washington University, Georgetown University, and the District of Columbia Fire and Emergency Medical Services Department. Dr. Dhindsa helped establish the emergency medical services and disaster medicine fellowship at George Washington and Georgetown Universities in Washington, District of Columbia, as well as the fellowship program at Virginia Commonwealth University. He currently teaches emergency medical services to multiple groups of learners, including field emergency medical services personnel, residents, fellows, and faculty.

Deborah DiazGranados, MS, PhD, is an assistant professor in the School of Medicine at Virginia Commonwealth University, in Richmond, Virginia. She earned her MS and PhD in industrial–organizational psychology from the University of Central Florida. Her over ten years of research and consulting experience has focused on teamwork, leadership, and team/leader effectiveness. Dr. DiazGranados's areas of expertise include leadership and team development, in particular in complex contexts such as healthcare, space teams, and the military. Her research is focused on understanding the influence of teamwork and leadership processes on individual- and team-level outcomes. Dr. DiazGranados's research has been funded and supported by the Department of Defense, the National Aeronautics and Space Administration, the National Institutes of Health, the National Center for Advancing Translational Science, the National Cancer Institute, and the National Science Foundation.

Aaron Dietz, PhD, is a human factors psychologist and faculty member at the Johns Hopkins University School of Medicine, Baltimore, Maryland, with dual appointments in the Armstrong Institute for Patient Safety and Quality and Department of Anesthesiology and Critical Care Medicine. Dr. Dietz has worked on an array of applied research projects dedicated to improving safety and performance that have required the application of state-of-the-art methodologies from human factors science and psychometric test development, such as team task analysis, survey and observational research, usability analysis, generalizability theory, and the application of sensor-based technology to measure teamwork and workload. His recent research has focused on understanding the drivers and barriers to team performance in critical care, developing an approach to measuring teamwork in critical care, and conducting studies to validate the psychometric properties of the measurement tool.

Stuart Donn, PhD, is a corporate education consultant with experience in emergency medicine systems. He received his PhD in educational assessment, evaluation, and research from Cornell University, Ithaca, New York. He has been responsible for curriculum design, program evaluation, and research projects incorporating cognitive task analysis for greater effectiveness of professional activities.

Alan W. Dow, MD, MSHA, is the Ruth and Seymour Perlin Professor of Medicine and Health Administration and director of the Center for Interprofessional Education and Collaborative Care at Virginia Commonwealth University, in Richmond, Virginia. Under his leadership, the center develops, implements, and studies initiatives in interprofessional education and collaborative practice across seven schools at Virginia Commonwealth University, the Virginia Commonwealth University Health System, and the surrounding community. He has been supported in this work with funding from the Josiah H. Macy Jr. Foundation as one of the inaugural class of Macy Faculty Scholars, a highly competitive national program focused on developing the next generation of educational leaders. He has also been funded by the Health Resources and Services Administration and the Centers for Medicare and Medicaid

Services. A practicing internist, Dr. Dow attended medical school at Washington University in St. Louis and completed residency in internal medicine and a health administration degree at Virginia Commonwealth University.

Keaton A. Fletcher, MA, is a doctoral student at the University of South Florida, Tampa, Florida, studying industrial–organizational psychology with a concentration on occupational health psychology, under the guidance of Dr. Wendy L. Bedwell. Prior to starting the doctoral program, Fletcher received his BS in neuroscience and BA in psychology from Washington and Lee University, where his academic focus was in health sciences. His research interests include occupational stressors such as cognitive workload; training design and evaluation; team training, dynamics, and performance; and work-related health outcomes. Specifically, Fletcher has focused on the negative effects of interruptions, the buffering effects of mindfulness, the ideal design of checklists to aid in training and performance, and both physiological and psychological strain and well-being outcomes. Currently, he is a member of the Society for Industrial and Organizational Psychology, SMA, Human Factors and Ergonomics Society, and INGroup. Fletcher is also an American Council on Exercise-certified wellness consultant and TeamSTEPPS master trainer.

Megan E. Gregory, PhD, is an Advanced Fellow in Educational Leadership at the Center for Innovations in Quality, Effectiveness and Safety at the Michael E. DeBakey Veterans Affairs Medical Center and Baylor College of Medicine in Houston, Texas. Her work focuses on developing, implementing, and evaluating educational programs and interventions in healthcare. She has particular expertise in quality improvement, program evaluation, teamwork, training, and statistical methods. Dr. Gregory obtained her PhD in industrial–organizational psychology from the University of Central Florida in 2015.

Shane E. Halse, MA, is a PhD student studying in the College of Information Sciences and Technology of Pennsylvania State University, University Park, Pennsylvania. Currently, his primary focus is in the domain of big data and crisis response, with a particular interest in improving the adoption of social media data in emergencies. Halse holds a master's degree in software engineering, which he utilizes to develop tools and application for improving crisis response managers' abilities to respond effectively to a disaster. In addition to his academic pursuits, Halse has been a member of a local search and rescue team, from which he has gained valuable insights into the world of emergency response.

David Hostler, MD, is a professor, the chair of the Department of Exercise and Nutrition Sciences, and a clinical professor of emergency medicine at the University at Buffalo, Buffalo, New York. He earned his BA and MS degrees from Wright State University and his doctorate in physiology from Ohio University. His research interests are human performance and physiological responses of public safety personnel working in protective equipment. With 29 years of experience in public safety, he is a founding faculty member and the director of the Emergency Responder Human Performance Lab and the director of the Center for Research and Education in Special

Environments. Dr. Hostler is an expert in performance in extreme environments and has been funded by the National Institutes of Health, the Federal Emergency Management Agency, and the Department of Defense to study prehospital care and performance among first responders and military personnel.

Ashley M. Hughes, PhD, is a researcher at the Center for Innovations in Quality, Effectiveness, and Safety, Michael E. DeBakey Veterans Affairs Medical Center, and the Baylor College of Medicine in Houston, Texas. She received her doctoral degree in applied experimental human factors from the psychology program at the University of Central Florida, with her masters in modeling and simulation. Her research focuses on teamwork, team training, and simulation primarily in medical settings.

Heather C. Lum, PhD, is an assistant professor at Penn State Erie, The Behrend College, Erie, Pennsylvania. She earned her PhD in applied experimental and human factors psychology from the University of Central Florida in 2011. Her primary research interest focuses on perceptions of technology, specifically the ways in which technology is impacting the way we interact with each other as humans. Other areas of interest include the use of psychophysiological measures such as eye tracking and vocal analyses to better determine and study the critical applied cognitive and experimental topics of interest such as spatial cognition and human–human and human–robot team interactions. She has also turned her attention to the use of games for training and educational purposes. In addition to her research pursuits, Lum is a faculty advisor of Psych Club, the local chapter of Psi Chi; the program chair of the Education Technical Group, and a former webmaster and newsletter editor for the Aug Cog Technical Group.

Brian J. Maguire, DrPH, is a professor in the School of Medical and Applied Sciences, Central Queensland University, Australia. He has a doctoral degree in public health from the George Washington University in Washington, District of Columbia, and a master's degree in health administration from Central Michigan University. He is a 2009 Senior Fulbright Scholar. Previously, he held the roles of clinical associate professor, graduate program director, and research center director for the University of Maryland, Baltimore County. For three years, he worked on a bioterrorism preparedness program for the US Department of Homeland Security. He has been a member of the Centers for Disease Control and Prevention's National Occupational Research Agenda and Public Safety Council as well as numerous National Highway Traffic Safety Administration emergency medical services committees, including the national emergency medical services safety committee and the Emergency Medical Services Agenda for the Future committee. For two decades, he worked in the New York City healthcare system as an administrator, educator, and paramedic.

Evan McHughes Palmer, PhD, is an assistant professor of psychology in the Human Factors Program at San José State University, San Jose, California. He earned his PhD in cognitive psychology from the University of California, Los Angeles, spent four years as a postdoctoral researcher at Harvard Medical School, and then was a faculty member of Wichita State University's Human Factors Psychology program for nine years. Dr. Palmer's research centers on applying principles from cognition, attention, and perception to solve real-world problems. Recent research topics in healthcare include doctors' mental representations of patient information for hand-offs, documentation workflow in a pediatric emergency department, and assessment and improvement of obstetrics and gynecology clerkship feedback for medical students. Dr. Palmer also researches how gamification techniques shape attention and motivation, web page perception, shape perception, visual attention, and visual search. He currently runs the Learning, Attention, Vision, and Application Lab at San José State University.

Dan Nathan-Roberts, PhD, is an assistant professor of industrial and systems engineering at San José State University, San Jose, California. Dr. Nathan-Roberts completed his PhD at the University of Michigan and an Agency for Healthcare Research and Quality postdoctoral fellowship at the University of Wisconsin–Madison. Dr. Nathan-Roberts has also been a fellow at the Food and Drug Administration, a physical ergonomics engineering researcher at the University of California Ergonomics Program, and an innovation consultant and has worked in the industry as a mechanical engineer and industrial engineer domestically and abroad. His research interests center on improving our understanding of the sociotechnical systems impacting healthcare providers and patients and quantifying the impact of holistic design decisions. Dr. Nathan-Roberts currently serves in volunteer positions in several engineering societies and is a champion of inclusion in engineering and of access to human factors education.

P. Daniel Patterson, MPH, PhD, is an assistant professor of emergency medicine at the University of Pittsburgh's School of Medicine, Pittsburgh, Pennsylvania. He earned his BS from Appalachian State University, MPH and PhD from the University of South Carolina, and MS from the University of Pittsburgh. He received postdoctoral and fellowship training in health services research and patient safety with support from the Agency for Healthcare Research and Quality and Society for Academic Emergency Medicine. He is trained in clinical research with support from the National Institutes of Health. His research interests include safety culture and teamwork, patient safety, clinician safety, fatigue and shift work, and safety of public safety personnel. Dr. Patterson is an expert in performance in extreme environments, and he has been funded by the National Institutes of Health, the Federal Emergency Management Agency, the Centers for Disease Control and Prevention/National Institute for Occupational Safety and Health, Department of Health and Human Services/Office of Rural Health Policy, and numerous foundations to study safety in the prehospital care setting. He is an active nationally registered paramedic.

Shawn Pruchnicki, MS, RPh, ATP, CFII, is currently a faculty member at The Ohio State University, Columbus, Ohio, where he teaches aviation safety, human factors, accident investigation, and complex aircraft operation at the Department of Aviation. His other Ohio State teaching responsibilities have included teaching courses in cognitive engineering at the Industrial Systems Engineering Department and clinical toxicology at the College of Pharmacy. Pruchnicki also owns a consulting company called Human Factors Investigation and Education and has performed work with many industries, including airlines, manufacturing, and fire departments, to name a few. During the first half of his career, he worked in the field of emergency medicine as a firefighter/paramedic with the Jackson Township Fire Department in central Ohio in addition to practicing as a pharmacist (toxicology) at the Central Ohio Poison Center. Prior to coming to The Ohio State University, he left emergency medicine and flew as a captain with Comair Airlines (Delta Connection) for ten years, flying the Canadair CL-65 regional jetliner. He has over 4500 hours of flight time, 3000 hours of which is in turbojets, and has a type rating in the CRJ, in addition to both airline transport pilot multiengine and commercial single engine certifications. He is also a certified flight instructor in basic, instrument, and multiengine operations and has given approximately 900 hours of flight instruction.

Michael A. Rosen, PhD, is a human factors psychologist and an associate professor at the Armstrong Institute for Patient Safety and Quality at the Johns Hopkins University School of Medicine, Baltimore, Maryland, with joint appointments in the School of Nursing and Bloomberg School of Public Health, Department of Health Policy and Management. Dr. Rosen is a human factors psychologist with research and applied experience in the areas of simulation, leadership, and team performance as they relate to patient safety and quality of care. He leads team training efforts in the Johns Hopkins Health System and has extensive experience in developing and validating team performance measurement systems incorporating observational, self-reporting, and novel sensor-based methods. He is recipient of the 2014 William C. Howell Early Career Award from the Human Factors and Ergonomics Society and a corecipient of the 2009 M. Scott Myers Applied Research in the Workplace Award from the Society for Industrial and Organizational Psychology.

Eduardo Salas, PhD, is a professor and the Allyn R. and Gladys M. Cline Chair in Psychology at Rice University, Houston, Texas. Previously, he was a trustee chair and a Pegasus Professor of Psychology at the University of Central Florida, where he also held an appointment as the program director of the Human Systems Integration Research Department at the Institute for Simulation and Training. Before joining the Institute for Simulation and Training, he was a senior research psychologist and the head of the Training Technology Development Branch of Naval Air Warfare Center Training Systems Division for 15 years. During this period, Dr. Salas served as a principal investigator for numerous research and development programs that focused on teamwork, team training, simulation-based training, decision-making under stress, safety culture. and performance assessment. Dr. Salas has coauthored over 450 journal articles and book chapters and has coedited 27 books. He is also the recipient of the 2012 Society for Human Resource Management Losey Lifetime Achievement Award and the 2012 Joseph E. McGrath Award for Lifetime Achievement.

Nastassia Savage, BA, is a doctoral student at the Department of Industrial– Organizational Psychology at Clemson University, Clemson, South Carolina. Her research interests involve organizational health psychology including stress and burnout, interactional justice, leadership, and teams, particularly in healthcare settings. She is currently involved in projects assessing these relationships in extreme environments funded by the National Aeronautics and Space Administration and healthcare organizations. She previously earned her BA in psychology with a minor in sociology at the University of Central Florida in 2011.

David Schuster, PhD, is an assistant professor of Psychology at San José State University, San Jose, California. He holds a PhD in psychology, specializing in applied experimental and human factors psychology, from the University of Central Florida. Dr. Schuster's research centers on understanding individual and shared cognition in complex environments. He has conducted research in domains such as aviation, transportation security training, and military human–robot interaction. His research interests include the cognitive aspects of cybersecurity, situation awareness in human–automation teams, the impact of video games on task performance, and perceptual training for real-world pattern recognition. In 2016, Dr. Schuster was awarded a National Science Foundation grant to investigate the cognitive factors of computer network defense. Dr. Schuster's work has appeared in *Ergonomics* and the *Journal of Management and Engineering Integration*. He is a member of the Human Factors and Ergonomics Society.

Marissa L. Shuffler, PhD, is an assistant professor of industrial–organizational psychology at Clemson University, Clemson, South Carolina. Her areas of expertise include team and leader training and development with an emphasis on high-risk and complex environments (e.g., virtual, distributed, high reliability). Dr. Shuffler has conducted qualitative and quantitative research aimed at understanding healthcare leadership issues, evaluating leadership and teamwork issues in distributed teams for the National Aeronautics and Space Administration, and designing training interventions to improve multiteam system functioning. She has conducted this and similar research for government, military, and industry, including the US Army Research Institute, the National Aeronautics and Space Administration, the National Science Foundation, the Center for Army Leadership, the Department of Homeland Security, the US Air Force, and the Greenville Health System. Her work to date includes an edited book on multiteam systems, over 45 publications, and over 100 presentations. She received her doctorate in industrial–organizational psychology from the University of Central Florida.

Shirley C. Sonesh, PhD, obtained her doctorate in organizational behavior at the A. B. Freeman School of Business at Tulane University, New Orleans, Louisiana, in 2012. While at Tulane University, Dr. Sonesh's research focused on expatriate knowledge transfer in multinational organizations. Currently, Dr. Sonesh is an adjunct professor at Tulane University and conducts research on the topics of patient safety, teamwork in healthcare, expatriation, coaching, training, and human automation systems. Dr. Sonesh has published papers in *Consulting Psychology Journal, Journal of Applied Psychology, Journal of Global Mobility, The Joint Commission on Quality and Patient Safety,* and *Families, Systems, & Health,* among others.

She has been invited to both national and international conferences to present her research related to these fields.

James L. Szalma, PhD, is an associate professor at the Psychology Department of the University of Central Florida, Orlando, Florida. He received a BS in chemistry from the University of Michigan in 1990 and an MA in applied experimental/human factors psychology in 1997 from the University of Cincinnati. He received a PhD in applied experimental/human factors psychology in 1999 from the University of Cincinnati. The theme for his laboratory, the Performance Research Laboratory, is how variations in task characteristics interact with the characteristics of the person (i.e., cognitive abilities, personality, emotion, motivation) to influence performance, workload, and stress of cognitively demanding tasks. His primary research interests include signal/threat detection (e.g., friend/foe identification), training for threat detection, fuzzy signal detection theory, and how the characteristics of tasks and operators interact to influence performance in the context of tasks that require sustained attention or that include human–automation interaction.

Matthew D. Weaver, MPH, PhD, is a research fellow at Brigham and Women's Hospital and Harvard Medical School, Boston, Massachusetts. He completed an undergraduate degree in emergency medicine, a master of public health degree, and a doctoral degree in epidemiology at the University of Pittsburgh while working as a paramedic in the Pittsburgh area. Dr. Weaver is interested in understanding the effects of sleep and circadian factors in settings with severe occupational demands. His current work involves identifying individual- and workplace-level factors amenable to intervention with the goal of improving occupational health and safety.

1 Introduction to Human Factors and Ergonomics of Emergency Medical Services

Paul Misasi and Joseph R. Keebler

CONTENTS

> Human error in medicine, and the adverse events that may follow, are problems of psychology and engineering, not of medicine.
>
> **John W. Senders**
> *Medical Devices, Medical Errors, & Medical Accidents, in Human Error in Medicine, 1994*

The delivery of emergency medical services (EMS), or what we will interchangeably refer to as *prehospital medicine* (arguably a more accurate moniker*), is a complex enterprise that exists at the confluence of medicine, public safety, public health, and transportation—both air and ground (Shrader, 2015). All but a handful of decades old, it continues to seek its place among the professional domains, especially in the United States. EMS agencies struggle for recognition as a distinct but critical element of the public's safety net and gateway into the healthcare system. Although EMS is generally well respected, it is often forgotten and receives substantially less funding and recognition than its law enforcement and fire suppression counterparts. In 2005, a New York University report investigating the preparedness of the United States' first responders revealed that EMS agencies (apart from fire departments) received less than 4% of federal monies made available in the aftermath of the devastating terrorist attacks of September 11, 2001. Unfortunately, this is a reality that many EMS providers and managers have become all too familiar with. However, EMS professionals continue to

---

* We recognize also that out-of-hospital medicine would probably be the best fit for the profession; but in aligning this book's vernacular, we chose the former.

forge ahead, and demonstrable progress has been made elevating the professionalism of the domain and the delivery of the service. As we write this chapter, a number of factors are beginning to converge that we believe will only increase the importance that EMS providers, managers, system designers, and leaders place on quality improvement, resilience, safety, and sustainability. Likewise, human factors, systems and industrial engineers, ergonomists, and patient safety researchers will find new and largely untapped opportunities that have been overlooked in the past.

Although certainly not an exhaustive list, we believe four interacting influences are worth mentioning as they bear on the importance of incorporating human factors in prehospital medicine and its systems of delivery. First, a significant leap in the industrialization of EMS was made in the 1980s to the 1990s through the work of Jack Stout at the University of Oklahoma's Center for Economic Management and Research, funded by the Kerr Foundation (Shrader, 2015). Since then, agencies have reconsidered the constraints and the opportunities afforded by system design with an economic lens. Stout's work identified the unique operating and market characteristics of ambulance services that influence the substantial variability in system cost compared to the quality of service, revealing critical elements of system design and operational strategies to optimize the efficiency and the effectiveness of service delivery models (Stout, 1994).

Secondly, the work of Edward Deming and others in the field of quality management has influenced general management theories and has proven the importance of systems thinking as it pertains to business processes and their design and improvement. Perla, Provost, and Parry (2013, p. 172) reviewed how the four tenets of Deming's System of Profound Knowledge are critical to healthcare's ability to make improvements:

- Appreciation for a system: A focus on how the parts of a process relate to one another to create a system with a specific aim
- Understanding variation: A distinction between variation that is an inherent part of the process and variation that is not typically part of the process or cause system
- Theory of knowledge: A concern for how people's view of what meaningful knowledge is impacts their learning and decision-making (epistemology)
- Psychology: Understanding how the interpersonal and social structures impact performance of a system or a process

As will become evident throughout this chapter and this book, there is a substantial overlap between the sciences of improvement and human factors across all four of these areas.

Thirdly, in 2012, the American Board of Medical Specialties approved EMS medicine as an official subspecialty of medicine, recognizing the unique environment, the distinct body of knowledge and skills necessary, and the constraints of care delivery, and the American Council for Graduate Medical Education subsequently began awarding fellowships in EMS.

Lastly and arguably the most influential in the United States, the Affordable Care Act of 2010 has begun to remodel payment incentives from a volume-based, fee-for-service standard to a quality of care paradigm. Although EMS has not been explicitly targeted by payers for a demonstration of value and quality as a contingency for reimbursement, leaders in the industry realize that the clock is ticking. Efforts to prepare for this eventuality are currently underway with an initiative known as *EMS Compass*, funded by the National Highway Traffic Safety Administration and administered by the National Association of State EMS Officials. One goal of EMS Compass is to develop the practical metrics by which EMS agencies can demonstrate quality and value before those metrics are developed and imposed by external agencies with less interest in their feasibility (National Association of State EMS Officials, 2015).

To summarize, as EMS rightfully aligns itself more closely with healthcare, the Affordable Care Act has communicated and codified *that* it is imperative to demonstrate quality and value of service; EMS Compass efforts that are underway will assist with the identification of *what* requires our clinical and managerial attentions, and through a grounded understanding of human factors and systems engineering, we can learn *how* to intervene in meaningful ways that do not rely on perfect knowledge, perfect skill, or perfect performance from providers at the "sharp end" to achieve improvement.

Ambulance services throughout Europe and Australia have been incorporating the analysis and the design tenets of *physical* ergonomics for some time and are leading the way in improving the physical safety of providers; meanwhile, their American counterparts are just beginning to update the decades-old standards and specifications for ambulance design to improve safety (see the study by the Commission on Accreditation of Ambulance Services, 2016). We all however, remain a long way from our goal. Maguire et al. (2014) recently published the findings of an Australian investigation, citing that the risk of serious occupational injury for paramedics was seven times higher than the national average and that the fatality rate was six times higher than the national average. Twelve years earlier, Maguire et al. (2002) published findings that the occupational fatality rate for paramedics in the United States was two and a half times higher than the national average.

The scientific investigation of the *cognitive* ergonomics of prehospital care has only begun to scratch the surface. Since the Institute of Medicine's report *To Err Is Human* (Kohn, Corrigan, and Donaldson, 1999) was released 17 years ago, the science of human factors has increasingly focused on applications in the medical domain—covering topics such as device design and user interface, medication safety, teamwork, effects of stress and fatigue, workflow, cognitive artifacts, expertise, social and organizational cultures, judgment and decision-making—all with the aim of understanding their effects on the safety and the quality of medical care. Efforts have been made to translate the practices developed in other complex sociotechnical domains such as aviation, but prehospital medicine has not gained the same amount of attention from the research community. Thus, this book begins as an endeavor to fill a niche of applying the science of human factors and ergonomics (HF/E) to prehospital medicine, with a hope that it will not only begin an important and continuing conversation between the two disciplines but also open the eyes of everyday practitioners, managers, EMS physicians, and leaders to a science that can aid them in their daily performance of work tasks, preserve and protect their livelihood, help save the lives of their patients, facilitate optimal performance as clinicians, and optimize processes to achieve organizational goals.

Modern healthcare has been called the most complex domain of work known to humans (Gluck, 2008), a statement that was referring to healthcare delivered in controlled settings (i.e., hospitals). EMS providers face these complexities and the additional factors of delivering healthcare in a mobile, volatile, unpredictable, and unforgiving environment. As such, they arguably have one of the most cognitively and physically demanding jobs in healthcare. It is medicine on the front lines, sometimes literally in the trenches. We believe that a book such as this will enhance the current best practices and the methods by which they are achieved. It is our hope that this work leads to new efforts and relationships between HF/E and EMS. With the eventual move of EMS to the Department of Health and Human Services in the United States, this book could help lay the groundwork for a long-lasting relationship between the two fields. This chapter will provide a high-level overview of what HF/E is, what it is not, and how it can and has been applied successfully.

WHAT IS HUMAN FACTORS?

HF/E is a broad, interdisciplinary science based on psychology and engineering that deals with the interface between humans and their work systems. Born out of necessity during World War II,* it integrates knowledge and practice from multiple disciplines, including but not limited to computer science, organizational management, cognitive and social science, ergonomics, engineering, and industrial design (Wickens et al., 2004, 2013). The goal of HF/E is to make systems safe, reliable, intuitive, and supportive of the human operator by enhancing their work experience and optimizing performance (Russ et al., 2013). HF/E science takes a multilevel approach that focuses on individuals, teams, tasks, tools, systems, and organizations to best understand the complexity of the work system. The term *ergonomics* in a pure, holistic sense is defined as the study of work that is broadly inclusive of how the human mind and body interact with information and tools in the environment to achieve the goals of work (Wickens et al., 2013). While this definition is more commonly understood among our European and Australian counterparts, in colloquial American parlance, the term more commonly refers to the design of physical environments, tools, and devices (e.g., chairs, grips, and seats) in a more anthropometric[†] sense. There are a number of terms that have become common in order to capture the nuances of specialized HF/E practice as the built environment becomes more complex and technical, just as there are various and increasing levels of specialization in medicine (e.g., physician → surgeon → plastic surgeon; physician → orthopedist → hand surgeon; physician → internist → cardiologist). Some examples of these terms include *cognitive ergonomics, cognitive engineering, cognitive systems engineering, macroergonomics, usability,* and others that are generally encompassed by the umbrella term of *human factors* (Lee and Kirlik, 2013). However, in HF/E, there is a substantial overlap among subspecialties, and they are not as conveniently distinct as organ systems in the human body or necessarily divided in terms of skill sets as they are in medicine. HF/E is considered a science because its practitioners have usually earned a PhD that requires the study and the application of rigorous scientific

---

* For a review of the historical development of human factors, see the study by Meister (1999).
† Anthropometry is the science that evaluates the metrics of human sizes, forms, and functional capabilities (Centers for Disease Control and Prevention, 2016).

methods and statistical evaluation for describing a phenomenon with a high degree of reliability and validity, beyond what may be attributable to chance.

Finally, the phrase *human factors approach* is commonly used to mean and is arguably synonymous with the phrase *systems approach*, in the sense that both seek to design systems that support human performance (while accounting for human limitations) and are resilient to perturbations or unanticipated events (Dekker, 2011).

Russ et al. (2013, p. 802) describe that there are two goals for the HF/E application in medicine: "1) support the cognitive and physical work of healthcare professionals and 2) promote high quality, safe care for patients." In relation to prehospital care, HF/E can aid providers, managers, and leaders by enhancing the design of their medication preparations and packaging, tools, gear, ambulances, communication systems, teamwork (from the dyadic, two-person interaction to the much larger multiteam systems), organizational policies, and general work practices. This book is organized to focus on the aspects of the individual, the team, and the organization to holistically capture the delivery of prehospital medicine.

WHAT HUMAN FACTORS IS NOT

When someone says that they are a physician, they have communicated to you that they have a professional doctorate degree in medicine (MD for allopathic physicians; DO for osteopathic physicians), which means that they have obtained the broad knowledge base about the physiology, the pathophysiology, and the treatment regimens of the human body that all physicians receive; what the term *physician* has not communicated is their particular specialty or which branch of medicine they have chosen to focus their practice on (e.g., family medicine, orthopedics, etc.). In the same manner, when someone indicates that they have a doctorate in psychology, it would be incorrect to assume that they are a clinical psychologist who sees patients in an office or a hospital performing talk therapy. This is a leap that is analogous to assuming that all physicians are surgeons, although the converse is certainly true—all surgeons are physicians. Psychology comes in many different flavors (the American Psychological Association recognizes 54 divisions), and along the same lines, engineering has four main categories, with many branches and divisions. Thus, when someone says that they are a human factors engineer (or human factors psychologist), they have told you that they have a broad base of knowledge and training in cognitive/social science as well as in industrial/systems engineering. What they have not told you is their particular specialty or interest(s) of study, for example, psychophysics, visual perception, computer interface/display design, or usability.

Russ et al. (2013) highlight some myths about HF/E so that the uninitiated are not misguided in their expectations. To emphasize, human factors is not just about soft skills (i.e., effective communication, teamwork, leadership, etc.), although they are indeed important areas of investigation and practice. Secondly, HF/E is not about manipulating or training people how to account for their "human factors" or eliminate human error. On the contrary, human factors seeks to optimize human performance for whatever particular situation, environment, tools, constraints, pressures, and goals people have. The point of human factors is to bring to bear our knowledge of people's physical and cognitive limitations in the design of work systems such that they are robust to the vulnerabilities of the human condition. A human factors expert will be the first person to tell you that the term *human error* is meaningless, as well as any of its synonyms: *loss of situation awareness, failure of crew resource management*, etc. (Dekker and Hollnagel, 2004). These terms are not an explanation of anything in and of themselves; they are the starting point for investigation. Additionally, human behavior is the least responsive element of a system to remedial intervention, especially if the context that influences behavior is ignored. In fact, it is precisely our human capacity for adaptation and innovation that prevents poorly designed systems from failing (Woods et al., 2012).

It is also imperative to recognize that training as a remedial management intervention has very specific utility, just as different medical therapies have specific indications and contraindications. In other words, not every problem or issue that arises throughout the course of work can be fixed by hauling everyone into a classroom. Thirdly, HF/E is not just about dials, knobs, and checklists. As healthcare (to include EMS) slowly becomes more informed about human factors, it is tempting to take ideas off a shelf and somehow insert them into practice (e.g., a checklist as a cognitive aid). Just as it has been said of prehospital care—"a competent paramedic knows how to intubate a patient, an expert paramedic knows when not to"—HF/E requires professionals with the depth of knowledge in the field to know when a particular intervention such as a cognitive artifact (to use a more precise term) is appropriate and when it is not; otherwise, interventions may be met with failure or limited/transient success. Finally, HF/E does not promote blame-free policies or organizational cultures; rather, it promotes the recognition that human behavior is inextricably bound and produced within a rich, situated context. As such, those responsible for designing systems contingent upon human performance must appreciate the influences that those systems impose upon performance, given human limitations.

A human factors approach balances personal accountability and managerial accountability for system design, which, perhaps given the historical extent to which punishment and individual-level remediation has been used to fix problems in healthcare, it may seem a bit blame-free at first. Reason (1994), a cognitive psychologist and the author of *Human Error* and many other works, has written time and time again that blame as a managerial tool for behavior modification is popular because it is fast, easy, and emotionally satisfying. It may also seem to get the desired results, but the truth is that a blame culture only impairs a system's ability to learn. System improvement, on the other hand, requires expertise, time, effort, capital, leadership, humility, and courage to expose the decisions, the process flaws, and the fundamental goal conflicts that seed latent error into the system.

HUMAN FACTORS HAS ALREADY MADE A DIFFERENCE IN EMS

An example of what human factors is capable of achieving in EMS, not just in terms of processes but also in terms of clinical outcomes and human lives, is the rapid spread of the "pit crew approaches" to field resuscitation of cardiac arrest patients, also known as high-performance cardiopulmonary resuscitation. Given the growing body of literature extolling the importance of uninterrupted chest compressions to improve the likelihood of restoring a patient's cardiac activity and spontaneous circulation while preserving the possibility of functional neurological outcomes, innovative methods have been employed to orchestrate the process of resuscitation. Elements such as predefined roles, communication protocols, callouts, space organization and utilization, cognitive artifacts, explicit multiteam coordination, expectancies, and others have been included in many designs. See Figures 1.1 through 1.3.

CARDIAC ARREST CHECKLIST   **R – 21**
REFERENCE   V1.2 (12-12-12)   Effective 11-1-2012

BLS [No pauses over 10 seconds]

    ☐ Designate and Announce BLS Team Leader (first name included) to all providers on scene upon arrival

Compressions

    ☐ Metronome (Ensure 110 per minute setting)
    ☐ SWITCH every 220 compressions to new compressor, NO EXCEPTIONS
    ☐ Determine who is next for compressions immediately after each switch and move into position

Cycle	1	2	3	★	4	5	6	7	8	9	10	11	12

Ventilations and Airway Management

    ☐ Non-Hypoxic Arrest | High flow O_2 by NonRebreather Mask/OP for first three (3) cycles of arrest (660 compressions)
    ☐ Hypoxic Arrest | High flow O_2 by BVM/OP for entire code with interposed ventilations at 6/minute (20:1)
    ☐ ★ Switch from CCR to CPR with BVM/OP after 660 compressions and interpose ventilations at 6/minute (20:1)
    ☐ Place $ETCO_2$ immediately upon availability

Defibrillation

    ☐ Attach AED and power on   **Shock** ☐   ☐   ☐   ☐
                                                                  1   2   3   4

ALS [Do not interfere with BLS triangle]

    ☐ Designate and Announce Code Commander (first name included) to all providers on scene upon arrival

ECG

    ☐ Verify monitor is in PADDLES mode at all times unless actively pacing
    ☐ Announce ECG interpretation and protocol for rhythm found on ECG

$ETCO_2$

    ☐ Place $ETCO_2$ filter line to BVM immediately
    ☐ Verify $ETCO_2$ waveform is present and being monitored
    ☐ Print $ETCO_2$ waveform [Upon airway placement | Any patient movement | Termination of care]

Airway

    ☐ After 660 compressions, place advanced airway if needed (Do not stop compressions for airway management)

Address Complications

    ☐ Assess blood glucose
    ☐ Gastric distention has been considered/addressed
    ☐ If unresolved or persistent arrest situation, look for and treat if indicated:
        ☐ Hypovolemia   ☐ Tension pneumothorax
        ☐ Hypo/Hyperkalemia   ☐ Toxins
        ☐ Hydrogen ion (Acidosis)   ☐ Thrombosis
        ☐ Hypoxia   ☐

ROSC

    ☐ 12 Lead ECG following ROSC
    ☐ Initiate chilled saline bolus if not already started intra-arrest per protocol

---

FIGURE 1.1 Cardiac arrest checklist as a cognitive artifact developed by and incorporated into the Wichita–Sedgwick County EMS System Medical Protocols. (Reprinted with permission from Medical Society of Sedgwick County, 2012.)

Cardiac arrest pit crew roles and responsibilities

Position #1

Role	Responsibilities
To facilitate continuous compressions in cardiac arrest and assist with airway/ventilation. Positioned at patient **right**. **Assigned** to fire fighter or paramedic on first in unit.	• Assesses unresponsiveness/pulselessness and initiate compressions • Alternates compression every 220 compressions with position #2 • Counts compressions in 20's and call out 17, 18, 19, 20 each time • Ventilates with BVM in off cycle (20:1) • Assists with airway management as needed

Position #2

Role	Responsibilities
To facilitate continuous compressions in cardiac arrest and assist with airway/ventilation. Positioned at patient **left**. **Assigned** to fire fighter or paramedic on first in unit.	• Brings and operates AED or LP 12 • Initiates metronome • Applies oxygen via NRM at high flow • Alternates compressions every 220 compressions with position #1 • Counts compressions in 20s and calls out 17, 18, 19, 20 each time • Ventilates with BVM in off cycle (20:1) • Assists with airway management as needed

Position #3

Role	Responsibilities
To facilitate airway patency and ventilations. Positioned at patient **head**. **Assigned** to fire fighter or paramedic on non-transporting response unit.	• Monitors and manages airway for duration of arrest to ensure patency. Reacts to problems • Calls out compressions in increments of 20 (20, 40, 60, 80, ... 220) • Assembles and applies all airway equipment except ETT • Applies BVM/OP at 660 compression mark with two handed seal on mask • Monitors EtCO$_2$ values and communicates with team

Note: Personnel can rotate in and out of positions 1, 2, and 3 as needed so long as this does **not** interfere with care or interrupt CPR.

Code team commander (CC)—paramedic in control of monitor

Role	Responsibilities
Ownership of all clinical care. Directs all ALS care in coordination with BLS team leader oversight of all BLS care. Holds the final responsibility for patient well being. Positioned outside BLS triangle near patient's **legs**. **Assigned** to paramedic on transport vehicle.	• Communicates/coordinates with team leader • Makes all patients treatment decisions • Rhythm analysis, interpretation, and application of electrical therapy • Assures accurate on-going documentation of care (written and LP 12) • Determines when patient is moved/transported • Supervises DNR and code cessation issues • Responsible for all clinical communications • Responsible for overall conduct of resuscitation • Owns any advanced airway interventions (ETT or combitube) • Overall documentation of care for entire call

BLS team leader (TL)

Role	Responsibilities
Responsible for all BLS efforts in cooperation with code team commander. Monitors clock and communicates times. Owns BLS care. Positioned outside BLS triangle. **Assigned** to EMT or higher provider.	• Works checklist and calls out times to code commander (6, 10, 15, 20 min) • Ensures great BLS, specifically a ≥90% compression fraction • Tracks compression cycles and ensures switching compressors every 220 • Tracks and calls for ventilations at the 600 compression mark (3rd cycle) • Ensures integrity of BLS triangle • Assists with airway setup if needed • Overall documentation of care for all BLS activities and assists CC with ALS documentation

FIGURE 1.2 Roles and responsibilities of pit crew resuscitation providers from the Wichita–Sedgwick County EMS System Medical Protocols. (Reprinted with permission from Medical Society of Sedgwick County, 2012.)

FIGURE 1.3 Pit crew spatial organization from the Wichita–Sedgwick County EMS System Medical Protocols. Locations of participants are indicated by circles. The basic life support (BLS) triangle indicates the area where the basic life support measures of continuous chest compressions are taking place and that this space is reserved for only that task, given its special importance.

The list below represents a number of observed and engineered human factors elements at work in a field resuscitation, and as you can see, none of them have their foundation in medicine; they are psychological constructs and/or engineering concepts.

The following are human factor constructs implicated and engineerable in pit crew resuscitation:

- Teamwork
- Design of clinical decision guidance
- Implicit/explicit coordination
- Cognitive artifacts
- Interpersonal communication
- Transactive memory systems
- Skills, rules, knowledge
- Alarms and technology
- Situation awareness
- Expertise

- Decision making under stress
- Dynamic fault management
- Callouts and cross-checks
- Social construction of reality
- Trust in automation
- Task switching
- Diffusion of responsibility
- Pattern recognition

Note that this list does not represent an exhaustive list.

A WORD ABOUT HUMAN ERROR

"The occasional human contribution to failure in complex systems occurs because the systems require an overwhelming human contribution for their safety" (Dekker, 2006, p. 194). Through decades of studying risk and safety in domains such as nuclear power generation and aviation, the science of HF/E has uncovered the idea that safety does not simply exist in complex systems; in essence, it is not an inherent property of the system; it is an emergent property of the interaction between humans and the system. The old view of error (Dekker, 2002) assumed that systems are inherently safe and that it is the human operators who are responsible for their demise. So when something goes wrong, all that is necessary is to find the defective part (i.e., the human) and begin the cycle of name, blame, shame, and retrain. The new view of human error acknowledges instead that complex systems are inherently intractable, dangerous, and risky and that the human serves as both a point of resilience (oftentimes striking a fine balance between irreconcilable goal and priority conflict), and the trip wire for the latent errors that they have inherited (Reason, 2000).

Furthermore, one cannot simply study the phenomenon of human error without studying normal human performance; they are inextricably linked, with only the distinction of their outcomes to tell the difference. Human error and correct/successful performance are two sides of the same coin (Reason, 1990). "Human error is not random"; it is systematic (Dekker, 2002, p. 61). To suggest that behaviors can be judged by the success or the failure of their outcomes is to say that the ends justify the means; analogously, we would consent to the countless numbers of people who drive their vehicles while intoxicated as long as they are somehow able to not destroy any people or property in the process.

Unless your agency is hiring sociopaths, your providers want to help people and do their jobs well. They do not choose to err when they could have otherwise chosen not to err. To err is not a choice (Dekker, 2006). Errors are not the result of negligent or ill intent by the provider; more often than not, they are the result of the intractable complexity of the system in which providers work—the interactions of information in the operators' situated context given their cognitive constraints and priority conflicts. Behavior always occurs in a context; it follows that understanding the context is crucial to understanding the subsequent behaviors—a concept referred to as the *local rationality principle* (Dekker, 2002), refined from the decision science concept of bounded rationality (March, 1978). People do their best given the constraints of their situation, expertise, and cognitive resources. Oftentimes, the outcome is good, but when it is bad, we tend to reframe their actions in hindsight with information that the person did not or could not have at the time. Healthcare providers, and especially EMS agents, are faced with achieving a balance between safety (theirs and their patients') and achieving the operational mission in a pressurized, unpredictable, patchwork system. Analyzing the systemic factors that contribute to errors, whether they be organizational policies, poorly designed technology, or inefficient task work, allows us to redesign the system to make it safer and more effective for our patients and our providers.

REFERENCES

Centers for Disease Control and Prevention, National Institute for Occupational Safety and Health. (2016). Anthropometry. Washington, DC: National Institute for Occupational Safety and Health. Retrieved from http://www.cdc.gov/niosh/topics/anthropometry/.

Commission on Accreditation of Ambulance Services. (2016). Ground vehicle standard for ambulances. Glenview, IL: Commission on Accreditation of Ambulance Services. Retrieved from http://www.groundvehiclestandard.org/wp-content/uploads/2016/03/CAAS _GVS_v_1_0_FinalwDates.pdf.

Dekker, S. (2002). *The Field Guide to Human Error Investigations*. Burlington, VT: Ashgate.

Dekker, S. (2006). *The Field Guide to Understanding Human Error*. Burlington, VT: Ashgate.

Dekker, S. (2011). *Patient Safety: A Human Factors Approach*. Boca Raton, FL: CRC Press.

Dekker, S., and Hollnagel, E. (2004). Human factors and folk models. *Cognition, Technology & Work*, 6(2), 79–86.

Gluck, P. A. (2008). Medical error theory. *Obstetrics and Gynecology Clinics of North America*, 35(1), 11–17, vii.

Kohn, L. T., Corrigan, J. M., and Donaldson, M. S. (Eds.). (1999). *To Err Is Human: Building a Safer Health System*. Washington, DC: National Academy Press.

Lee, J. D., and Kirlik, A. (Eds.). (2013). *The Oxford Handbook of Cognitive Engineering*. New York: Oxford University Press.

Maguire, B. J., Hunting, K. L., Smith, G. S., and Levick, N. R. (2002). Occupational fatalities in emergency medical services: A hidden crisis. *Annals of Emergency Medicine*, 40(6), 625–632.

Maguire, B. J., O'Meara, P. F., Brightwell, R. F., O'Neill, B. J., and Fitzgerald, G. J. (2014). Occupational injury risk among Australian paramedics: An analysis of national data. *Medical Journal of Australia*, 200(8), 477–480.

March, J. G. (1978). Bounded rationality, ambiguity, and the engineering of choice. *The Bell Journal of Economics*, 9(2), 587–608.

Meister, D. (1999). *The History of Human Factors and Ergonomics*. Mahwah, NJ: Lawrence Erlbaum Associates.

National Association of State EMS Officials. (2015). EMS Compass: Improving systems of care through meaningful measures. Falls Church, VA: National Association of State EMS Officials. Retrieved from http://www.emscompass.org/about-ems-compass/.

New York University, Center for Catastrophe Preparedness and Response. (2005). Emergency medical services: The forgotten first responder. New York, New York University. Retrieved from http://www.nyu.edu/ccpr/NYUEMSreport.pdf.

Perla, R. J., Provost, L. P., and Parry, G. J. (2013). Seven propositions of the science of improvement: Exploring foundations. *Quality Management in Health Care*, 22(3), 170–186.

Reason, J. (1990). *Human Error*. New York: Cambridge University Press.

Reason, J. (2000). Human error: Models and management. *British Medical Journal*, 320(3), 768–770.

Reason, J. T. (1994). Forward. In M. S. Bogner (Ed.), *Human Error in Medicine* (pp. vii–xv). Hillsdale, NJ: Lawrence Erlbaum Associates.

Russ, A. L., Fairbanks, R. J., Karsh, B.-T., Militello, L. G., Saleem, J. J., and Wears, R. L. (2013). The science of human factors: Separating faction from fiction. *BMJ Quality and Safety*, 22, 802–808.

Senders, J. W. (1994). Medical devices, medical errors, and medical accidents. In M. S. Bogner (Ed.), *Human Error in Medicine* (pp. 159–177). Hillsdale, NJ: Lawrence Erlbaum Associates.

Shrader, D. A. (2015). The ambulance market. In National Emergency Medical Service Management Association (NEMSMA) (Ed.), *Management of Ambulance Services* (first ed., pp. 79–98). Boston: Pearson.

Stout, J. (1994). System financing. In W. Roush (Ed.), *Principles of EMS Systems* (second ed., pp. 451–473). Irving, TX: American College of Emergency Physicians.

Wickens, C. D., Hollands, J. G., Banbury, S., and Parasuraman, R. (2013). *Engineering Psychology and Human Performance* (fourth ed.). Upper Saddle River, NJ: Pearson.

Wickens, C. D., Lee, J. D., Liu, Y., and Gordon Becker, S. E. (2004). *An Introduction to Human Factors Engineering* (second ed.). Upper Saddle River, NJ: Pearson Prentice Hall.

Woods, D. D., Dekker, S., Cook, R., Johannesen, L., and Sarter, N. (2012). *Behind Human Error* (second ed.). Burlington, VT: Ashgate.

2 Cognitive Factors in Emergency Medical Services

Evan McHughes Palmer

CONTENTS

A call comes in from a dispatch to the emergency medical services (EMS) post: "57-year-old female, unconscious." The paramedics jump in the truck, leave the station, and head into traffic with lights on and the sirens blaring. The driver negotiates traffic as quickly as possible, sometimes driving on the left side of the road, sometimes on the right, proceeding through red lights while vigilantly scanning to make sure traffic is yielding to the ambulance. The partner pulls up the dispatch information, reads a description of the situation, and then switches over to a map to help navigate to the call location. Upon arriving at the scene, the paramedics are met at the door by the patient's husband, who leads the team up to the second floor bedroom. By this time, the fire department's first responders are also on the scene and five emergency personnel crowd into the small bedroom with the husband. The patient is not unconscious after all, but is lethargic and unable to rise from the bed. Her skin is cold and clammy to the touch, her heart rate is less than 50 beats per minute, and she reports feeling nauseous and faint. The lead paramedic on this call starts asking her questions. "What have you eaten today? How much have you had to drink? Are you on any medications?" Through the line of questioning, a picture starts to emerge about the situation. The paramedic is concerned about the slow heart rate, but the patient assures him that her heart rate is always slow. The team administers antinausea medication by estimating the weight of the patient and choosing the right dose after performing complicated mental calculations, taking into consideration her age and comorbidities that may affect the medication's safety. The medication dosage is cross-checked between the team members to make sure that they have calculated the dosage correctly. The team also starts to administer intravenous (IV) fluid since they have determined that one of the patient's problems is that she is severely dehydrated. After a few minutes, the patient is well enough to stand and is loaded into a chairlift for transport down the stairs and out to the ambulance. She gets out of the stair chair and onto the stretcher under her own power. On the way to the emergency room, the paramedic in the back with the patient continues to administer fluids and gives another round of antinausea medications, again after calculating the correct dosage. He hooks up a 12-lead electrocardiograph (EKG) to the patient to assess for myocardial ischemia or injury as a step in his process of confirming or disconfirming his suspected diagnostic hypothesis. He switches between entering patient information into a laptop, radioing ahead to the emergency room, and keeping up a dialogue with the patient to better understand her medical situation. Once at the hospital, the patient is transferred to the emergency department staff and the paramedic recaps all that he learned about the medical situation and the aid he provided up to that point. Finally, with the call over, the paramedics restock the truck and prepare for their next call.

The above scenario is typical of an emergency medical call and involves numerous cognitive activities. While driving, the paramedics must perceive the traffic situation, attend to the motion trajectories of the cars, remain vigilant for cars not yielding to the ambulance sirens, and search for street signs leading to the call location. All the while, the driver is being briefed on the upcoming call by his partner, who is multitasking by flipping back and forth between the call ticket and the city map on the laptop while also managing radio communication on the way to the scene. The paramedics form a set of expectations about how the call will go based on the description from the dispatch and their years of experience. Such expectations help them prepare for what they are about to face, but may also have an undue influence on their decision-making later. After arriving on the scene and evaluating for threats to personal safety, the paramedics must quickly assess the situation and determine which parts of the call description were accurate and which were not. They must be aware of not only their environment but also the visual symptoms of the patient—her complexion, coloring, clarity of speech, and movements. The paramedics must not be too influenced by the original description of the patient's illness from the dispatch lest it causes them to overlook important clues or head down the wrong diagnostic path. Rather, they ask questions to form a more detailed mental model of the patient's illness so that they can quickly stabilize her and prepare her for transport. In this system, medication doses must be calculated and cross-checked with other team members to avoid mental math errors. (The cross-check verification procedure is unusual at the national level, but has reduced medication errors in this county [Misasi, Lazzara & Keebler, 2014].) A paper printout of the patient's EKG is examined for telltale patterns indicating a heart attack, a visual pattern recognition task. Details of the patient's situation must be remembered to report to the hospital and to write up in the patient care report, often hours after the call is over—which is taxing on working memory. The paramedics must remember which supplies they used during the call so that they can restock at the hospital after the call is over. All the while, the emergency medical team must have a good bedside manner and maintain a calm presence to reassure the patient that everything will be OK—even if this is the end of their shift, they are fatigued, and the patient is uncooperative or abusive.

Prehospital medicine is delivered in fast-paced, high-stakes environments in which cognitive and perceptual factors may determine the difference between life and death. Paramedics and emergency medical technicians assess a scene upon arrival and make quick diagnostic decisions that need to be accurate and hold up to scrutiny. Misreading a medication label, miscalculating a medication dose, or misdiagnosing a medical problem can all result in adverse outcomes for a patient. The domains of these types of errors—perception, attention, memory, and decision-making—are all studied by cognitive psychologists. Findings from these and other domains in cognitive psychology are relevant to practitioners and may help them better understand the mental demands of emergency medicine. The purpose of this chapter is to provide an overview of the field of cognitive psychology, talk briefly about its origins and major concepts, and describe how they may relate to prehospital medicine.

HUMANS AS LIMITED INFORMATION PROCESSORS

The field of cognitive psychology started in the 1950s and 1960s in an attempt to explain human mental phenomena such as problem-solving, language, memory, perception, attention, and reasoning (Gardner, 1985). At that time, the behaviorist approach was the dominant theoretical perspective in psychology and limited experimental investigations to only observable data, such as clearly identifiable stimuli and responses. This theoretical perspective constrained the sorts of topics that could be studied, and psychologists grew frustrated with behaviorism's inability and unwillingness to address the complexities of human thought, which is not directly observable. At the same time, psychologists were compiling a number of findings that could not be explained using the tools of behaviorism, setting the stage for a revolution in the field of psychology (Neisser, 2014). The cognitive revolution used the newest technology of the day—computers—as a way to help conceptualize the machinations of the human mind and became wildly successful by doing so (Gigerenzer & Goldstein, 1996).

The core of the cognitive approach is conceptualizing humans as information processors. Information comes into the human mind through the senses, is represented and processed in myriad ways, and then is transformed into thoughts, memories, decisions, or behaviors performed by the individual (Neisser, 2014). Furthermore, information is interpreted by the individual, based on their knowledge, expertise, and life experiences. Much of cognition is an active process, in terms of both acquiring information from the world and thinking about that information. A person might observe a certain set of symptoms in a patient, search for other related symptoms, think about what those symptoms mean, come to a conclusion about the diagnosis, and then act on that diagnosis to administer proper aid. Such behaviors impact the world, are observed perceptually, are interpreted cognitively, and lead to further behaviors that impact the world, and then the cycle repeats.

Cognitive psychology further recognizes that the mind is a *limited-capacity* information-processing system (Miller, 1956; Tsotsos, 1990). Why limited in capacity? One way to understand this is to consider that the world contains an essentially unlimited amount of information but the human mind has only a limited number of neurons. Therefore, it is computationally *impossible* for humans to process all of the information available to them (Tsotsos, 1990). Consequently, the human mind must take shortcuts when processing information, such as selecting only certain streams of information to process instead of others or making educated guesses rather than laborious calculations. In support of this notion, Miller (1956) famously proposed that humans are limited to about 7 ± 2 "chunks" of information that they can maintain in working memory at any given time (where a chunk can be thought of as a group of tightly associated pieces of information, such as the single term *EMS* rather than the three words *emergency*, *medical*, and *services*). Broadbent (1958) proposed that we limit the amount of information we process at any given time by filtering out irrelevant bits. By accepting processing limitations of humans as a given, cognitive psychology concentrated on how humans overcome those limitations through various mental strategies such as chunking and filtering. Such a perspective gives insights into not only what humans cannot do but also what they can do by using efficient strategies. Despite their limited cognitive resources, humans typically perform remarkably well in complex tasks such as prehospital medicine, mainly because they have adopted strategies to make the most use of the information that they can process.

The field of human factors adopted the notion of humans as limited-capacity information processors capable of making mistakes and added a systems perspective, which conceptualizes work environments as complex, interacting systems instead of collections of individuals (e.g., Dekker, 2014; Holden et al., 2013). Given that humans make mistakes, it is important to have resilient and redundant systems to catch those mistakes and to frame information and tasks in ways that humans process more naturally. User-centered design, information visualization, and cognitive engineering all depend on knowing the human mind's capacities and limitations so that technologies and work systems can be designed to minimize errors and maximize performance.

For a long time, the expectation in the medical community was that providers should deliver care with no mistakes. When mistakes did occur, especially when they resulted in injury or death, the individual care provider was held to account and punished for the transgression (the "bad apple" theory; Dekker, 2014). Much of this dynamics is still present in healthcare today, although thankfully a more system-level approach is being adopted by the community (e.g., Carayon et al., 2006; Holden et al., 2013). Under the systems perspective, when an error occurs, it is more useful to examine the overall situation and environment surrounding the incident to determine what went wrong (Strauch, 2002), rather than to blame the healthcare provider and stop there. The notion is that nobody sets out to harm a patient, so if an error happened, it must have seemed like the logical action to take at the time, given the information that was available to the worker. Therefore, errors must be understood within the context of the information available and decisions that healthcare workers must make (Dekker, 2014).

A watershed moment for the healthcare community in terms of how to conceptualize and correct medical errors happened when the Institute of Medicine issued its famous report *To Err Is Human* (Kohn, Corrigan & Donaldson, 2000). The report documents the vast number of medical errors that actually occur nationwide and proposes various steps to correct them. More importantly, the report represents a shift by the medical community toward conceptualizing medical providers as limited-capacity information processors and recognizing that medical errors will happen, especially when redundant error-checking systems are lacking.

While the healthcare field has just started to conceptualize humans as error-prone organisms, the field of cognitive psychology has considered humans to be limited information processors from its very founding (e.g., Miller, 1956). Such a perspective may be useful for the field of medicine in general and EMS in particular since cognitive psychologists have been cataloging the variety and extent of humans' mental limitations for over half a century.

Overall, the cognitive approach conceptualizes humans as having finite mental resources that can lead to errors under the right (or wrong) circumstances. The question is not whether humans will commit errors—they will—but rather, *why* do people make the errors they do? On the other hand, what mental operations support accurate performance and decision-making? Understanding the limitations and boundaries of human cognition and performance allows us to make systems that are robust to the errors that humans inevitably commit. Once the inevitability of human error is accepted, the focus then turns to both understanding when such errors will happen and building systems and processes that identify and correct potential errors before they happen. To understand when errors will occur, we must first understand the limits of human cognition.

COGNITIVE FACTORS IN EMERGENCY MEDICINE

The following sections offer a brief overview of some domains of cognitive psychology that are relevant to human factors in emergency medicine. Although they are not an exhaustive list, the topics discussed should provide some insights into the various ways that cognitive psychology topics may be applied to EMS scenarios.

ATTENTION

As mentioned previously, there is more information available in the world than we can process with our limited cognitive resources. Therefore, some information must be processed at the expense of the rest (Broadbent, 1958). Such differential processing is *attention*. The processed items are attended; the unprocessed items are ignored. Knowing which things to pay attention to and which can be safely ignored is a key element of expertise (Ericsson & Lehman, 1996).

There are many ways in which attention is used in emergency medicine. Even the most basic task of restocking a vehicle at the beginning of a shift involves a visual search for items either low in inventory or out of stock. The worker knows what the items look like and where they belong, and they direct their visual attention toward those aspects of the truck that they are restocking. When driving to a scene under lights and sirens, one needs to monitor traffic and shift their attention around to ensure that other vehicles are not going to come through the intersection.

When examining a patient, a paramedic directs their attention toward their instruments or the patient's body to gather information for a diagnosis. They might look at the patient's coloring, pupils, and heart rate or look for signs of wounds or cardiac arrest. Each of these tasks involves selecting a subset of the world to process while withdrawing processing of the rest of the world. Visual searches of these sorts involve systematic deployments of attention to stimuli in the world to gather information and form a mental representation of the situation (Custers, Regehr & Norman, 1996).

One interesting phenomenon related to attention and search is known as *satisfaction of search*. When performing a visual search for a target, there is a tendency for one to stop searching after a single target is found and therefore miss other targets that might be relevant (Berbaum et al., 1990; Tuddenham, 1962). In one famous example, radiologists examining a chest X-ray were able to find the tumor but failed to notice that the clavicle had been removed from the image (Potchen, 2006). Therefore, when assessing a situation, it is important to remember that just because one problem has been identified, that does not mean that there are no other problems that should be addressed. It is important to be thorough when arriving on scene and assessing a patient's injuries.

Another important aspect of attention for EMS is multitasking, which is having two (or more) active tasks you are trying to accomplish at once (Poposki & Oswald, 2010). During such occasions, we split our attention between the tasks—or, more accurately, we rapidly shift our attention back and forth between the tasks (Pashler, 1994). When driving to a scene, one must not only drive but also negotiate traffic, monitor radio chatter, strategize with one's partner, and navigate to the call location. When transporting a patient, the paramedic in the back of the truck must attend to the patient, accurately dose and deliver drugs, monitor vitals, and coordinate with the hospital. Since attention is split or constantly redirected while multitasking, it is easy to make mistakes. Some multitasking is unavoidable in EMS, but whenever possible, tasks should be completed one at a time with full attention to avoid the possibility of medical errors.

Schemas

Since humans cannot process all of the information available in the world, they usually do not even try. Instead, they look for patterns in the environment and draw conclusions from partial evidence. Some cognitive psychologists talk about this sort of pattern processing in terms of *schemas*, which are structured knowledge representations with default assumptions built in (Bartlett, 1932). For instance, if your friend tells you that they recently adopted a golden retriever, you instantly make a number of assumptions about their new dog without being consciously aware of doing so.

For instance, you may assume that the dog has four legs, long yellow hair, and a pleasant disposition. You have expectations about the size of the dog, its intelligence, the amount of grooming it requires, and whether or not it would play fetch. The point here is that we are constantly making assumptions and filling in information about objects, people, and events that we actually know very little about.

Schemas come into play when EMS personnel respond to a call. The dispatch gives a description of the situation, which may lead to assumptions about how the call will go. However, until EMS arrives on the scene, it is not clear what is really happening. If the description of the patient's symptoms is inaccurate, workers may go down the wrong diagnostic path due to assumptions from the dispatch's description. For instance, if you are told that you are heading to a scene to treat a 55-year-old male experiencing chest pains and shortness of breath, you might automatically assume that it has something to do with their heart when it could instead be an allergic reaction or adverse drug interaction. Similarly, when treating a patient in their early 20s who is having slurred speech and weakness on the right side of their body, you might be less willing to diagnose them as having a stroke as you would if they were in their 70s. The assumptions we make is that older people tend to have strokes, not younger people. While this is typically true, it is not always true, and it is important to be aware of any hidden assumptions one is making while treating a patient.

MEMORY

Memory is one of the most widely studied and well-understood topics in cognitive psychology. Many people tend to think of memory as a single resource—like a hard drive or a digital recorder—that stores experiences in a way that can be faithfully read out later. In reality, there are many types of memory and they work almost nothing like digital recording devices do. Memory is selective, biased, associative, and fragile. Without extensive training and practice in memory techniques, we have very little control of what we remember, as anyone knows who has forgotten the name of somebody they have just met. The following section will review several major types of memory resources, briefly describe their capacities and limitations, and then talk about the aspects of memory relevant to EMS.

The first distinction cognitive psychologists make is between short-term and long-term memory. *Short-term memory* contains the current items one is thinking about at any given time, which must be rehearsed to be maintained and can be easily lost (Baddeley, 2003). *Long-term memory*, on the other hand, is a vast storehouse containing events, facts, and skills, which is much more robust to loss and interference than short-term memory (Schacter & Tulving, 1994). Interestingly, items in long-term memory may be stored but momentarily irretrievable, depending on the current context and cues available (Bjork & Bjork, 1992). Understanding the nature and limits of human memory can help EMS workers avoid errors and maximize performance.

Short-Term/Working Memory

Short-term memory contains the conscious thoughts a person is currently considering. The terms *working memory* and *short-term memory* are often used interchangeably, although the former implies manipulation of items, whereas the latter does not necessarily. With regard to working memory, Baddeley (2003) proposed that there are several subsystems: a *visuospatial sketch pad* for mental imagery, a *phonological loop* for verbal items, an *episodic buffer* for remembering the temporal ordering of events, and a *central executive* for managing interactions between the previous three systems. An EMS worker may use the visuospatial sketch pad when imagining how to position a patient on a gurney or thinking about where to place equipment for optimal use. The phonological loop may be used to repeat a list of items mentally so that they can be maintained long enough to be written down. The episodic buffer may be used when listening to a patient describe the series of events that led up to their injury, which may be reported in a different order than they were actually experienced and thus need to be rearranged to form a proper narrative. The central executive, meanwhile, coordinates all of these activities and determines which working memory resources should be used when and to what effect.

A major feature of short-term/working memory is that it is limited in capacity, meaning that only a certain number of items can be maintained at once. Miller (1956) famously estimated the human memory span to be 7 ± 2 items, and while there have been different estimates over the years, the general notion that short-term memory has a limited capacity is widely accepted. One of the consequences of the limited nature of short-term memory is that items need to be transferred to a more permanent form of storage—either written down or transferred to long-term memory—to prevent them from being lost.

Long-Term Memory

Whereas short-term memory is limited in capacity, long-term memory is essentially unlimited and, like a scaffolding, the more knowledge one has in a certain domain, the more they have the capacity to learn (Bjork & Bjork, 1992). Long-term memory can be broadly divided into *declarative* and *nondeclarative* types (Schacter & Tulving, 1994). Declarative memories, sometimes called *explicit memories*, are the sorts of things you can say out loud to another person, such as what you ate for lunch or the capital city of France. Declarative memory can be further subdivided into *episodic* and *semantic types* (Squire & Zola, 1996). Episodic memories are memories of previous events that you have experienced and include information about the particular time and place at which the event happened. For instance, writing a narrative description of an emergency call after the fact relies heavily on episodic memory for the places and events experienced. Semantic memories, on the other hand, are memories of facts that you know, but which are not tied to a particular time and place. For instance, knowing that there are four chambers in the human heart or 12 pairs of cranial nerves does not depend on remembering the particular day in school that you learned that information.

Another interesting form of explicit long-term memory that cognitive psychologists have studied is *prospective memory*, which is the intention to remember to do something later (Brandimonte, Einstein & McDaniel, 2014). There are many tasks in EMS that require prospective memory. For instance, there is not always time to write down patient information, especially in emergencies, so one may make a mental note to tell the hospital about a patient's drug allergy. If certain supplies are used up on a call, you might have to remember to restock before the next call. If there is a delay in treatment because another treatment needs to be executed first, then remembering to do the original treatment is a prospective memory. In all of these cases, there is the intention to remember something later, which may or may not happen. Some of the keys to successful prospective memory retrieval are having both distinctive and specific retrieval cues at a later date, which can help recall one's original intentions (McDaniel & Einstein, 2000).

Nondeclarative memory, also called *implicit memory*, stores information that is harder to verbalize (Schacter & Tulving, 1994). For instance, knowing how to play an instrument, drive a car, or insert an IV tube with a gentle touch are all motor-based skills that are difficult to learn and hard to explain without mentally or physically simulating the process. Another interesting form of nondeclarative memories is habits, which are learned sequences of behavior that do not require explicit recall to perform (Squire & Zola, 1996). Anyone who has left the house to run an errand and accidentally started to drive to work because their mind was wandering can attest to the power of habits. In EMS and all safety-critical fields, it is important to establish good habits so that the correct procedures will be followed even in the face of distraction. Whether it is taking blood pressure, starting an IV, or hooking up equipment, the more times you have rehearsed doing something the smoother that the process goes. Furthermore, being able to perform the physical tasks of EMS smoothly reduces cognitive load, which in turn frees up mental resources for other items. Considering the fundamentally limited nature of human information processing, it is important to reduce cognitive load whenever possible.

Once way to improve memory and cognition is to off-load those processes onto physical artifacts that support decision-making. Specifically, procedures and cognitive aids (e.g., checklists) can help ensure that processes are being done correctly and relevant information is being considered (Gawande, 2010). They provide structure to support reasoning and alleviate cognitive workload, especially under times of stress and limited cognitive resources. A second way to improve performance in the field is to practice procedures by using high-fidelity simulation (Salas et al., 2008). Such exercises create opportunities for implicit habits to be learned in realistic environments and explicit recall to be aided using the memory cues provided by the simulation.

JUDGMENT AND DECISION-MAKING

Many of the mentioned examples have to do with gathering information and making decisions in the face of uncertainty. The branch of cognitive psychology that studies such processes is known as judgment and decision-making. It is probably not too controversial to say that humans do not always reason objectively and rationally about information. We do not think through all possible courses of action, weigh them, and then choose the optimal path in a dispassionate, systematic way (Klein, 2008). Rather, we typically use shortcuts in thinking—heuristics—to help us make decisions and we have biases in the ways that we consider information (Kahneman, 2011). While these heuristics and biases often work, they are not guaranteed to work and can cause humans to make poor decisions (Kahneman, Slovic & Tversky, 1982). Furthermore, the less time we have or the more stress we are under while making a decision, the less deliberative we are in the process (Cannon-Bowers & Salas, 1998). There are several heuristics and biases that are relevant to EMS and can impair optimal decision-making.

Anchoring and Adjustment

One heuristic people use when estimating quantities and magnitudes is *anchoring and adjustment*. When estimating some quantity, we tend to become anchored to an initial number and then adjust our estimate away from the anchor. This seems like a reasonable strategy except for two problems: (1) we tend to underadjust away from an anchor (Slovic & Lichtenstein, 1971) and (2) just about anything can serve as an anchor, whether it is relevant to the situation or not (Tversky & Kahneman, 1974). For instance, Tversky and Kahneman (1974) had participants spin a roulette wheel that had been rigged to stop at either 10 or 65. After spinning the wheel and having it land on a low or a high number, the participants were asked to estimate the percentage of African countries in the United Nations. People who spun a low number estimated a significantly lower percentage of African countries in the United Nations (25%) than did those who spun a high number (45%). What did the numbers on the wheel have to do with estimating the number of African countries in the United Nations? Nothing! But the mere fact that an arbitrary number was in the participants' heads when they made their estimate was enough to significantly influence their estimates.

In the realm of EMS, it is important to examine the anchors one is using when reasoning about quantities and magnitudes. For instance, if you are estimating the weight of a patient to calculate a drug dose, an initial incorrect guess about the weight might have a profound influence on the final weight used in the dosage calculation. When asking a patient how many pills they have consumed or drinks they have had, offering an initial estimate may inadvertently bias their report (e.g., "How many pills did you take? 10?"). It is important to be aware of the anchoring and adjustment bias in order to guard against it.

Availability

Another bias that can influence decision-making in EMS is *availability*, in which examples that more easily come to mind are judged to be more frequent. Tversky and Kahneman (1974) reported an example in which groups of participants heard a list of male and female names, some of whom were famous (e.g., Richard Nixon) and some of whom were less so (e.g., Lana Turner). One group heard a list with more famous male names and another heard a list with more famous female names. The participants were then asked to recall whether the list contained more men or more women. Despite the lists being the same length, participants judged the lists to contain more names of the sex that was more famous. The ease with which the famous male or female names came to mind during the recall phase caused the participants to think that there were more males or females on the list, respectively.

In the realm of EMS, if by happenstance a worker has experienced several rare cases in a row, she might be biased toward thinking that those cases are more common than they really are, which could affect her diagnostic decisions. Or, for example, if the city is in the middle of a heat wave and it is being reported all over the news, then one might be biased toward thinking that a patient who is nauseous and faint is experiencing heat stroke rather than some other cause. However, it would be prudent to consider other symptoms that could *disprove* the diagnosis of heat stroke rather than just those symptoms that are consistent with it, which brings us to our next decision-making bias.

Confirmation Bias

Confirmation bias is the tendency to consider only information that is consistent with a hypothesis or one's point of view rather than also considering information that might disprove it (Baron, 2000). In a famous example by Shafir (1993), two groups of participants evaluated a hypothetical scenario in which two parents were suing for the sole custody of their child. Both groups read the same descriptions of the parents: parent A had average income, health, and working hours; an average relationship with the child; and a relatively stable social life, while parent B had an above-average income, minor health problems, long working hours, an extremely close relationship with the child, and an extremely active social life. When one group of participants was asked, "To which parent would you *award* sole custody?" they picked parent B 64% of the time. Interestingly, however, when the other group was asked, "To which parent would you *deny* sole custody?" they also tended to pick parent B 55% of the time. In other words, Parent B was both awarded and denied custody of the child, depending on how the question was framed. Why would this preference reversal happen? When the participants considered reasons to award custody, they tended to concentrate on the positive aspects of parent B to confirm their decision, but when they considered reasons to deny custody, they concentrated on the negative aspects of parent B in the same vein. In other words, the participants tended to concentrate on information that was consistent with the decision they needed to make, rather than considering all of the information as a whole.

The confirmation bias reflects the tendency of people to frame their evaluation of information with regard to their goals or current hypotheses. People tend to seek out confirming information for their points of view and do not consider disconfirming information as readily. In the realm of EMS, the way a call is first described by the dispatch may influence the diagnostic decisions of the medical team. There may be a bias to frame one's thinking relative to the information reported rather than by considering other alternatives.

Due to many reasons already listed earlier—the dispatch's description of a call, recent calls that one has experienced, and assumptions about people based on their age, appearance, etc.—one might have a quick hunch as to the diagnosis of a patient upon arriving on the scene. The danger of the confirmation bias is that the EMS worker might only seek out information that would confirm their current hunch rather than also seek out information that might disconfirm it. It is important to not be locked into a conclusion quickly and to consider alternative explanations of how a patient presents.

CONCLUSION

The cognitive perspective is useful for thinking about emergency services and prehospital medicine. It recognizes that humans are limited in their information-processing abilities and biased in the ways they consider information. It recognizes that humans will make errors. Cognitive psychology can help to understand the information processing demands of tasks and the thinking and cognitive resources available in emergency medicine and can help to specify the sorts of measures and countermeasures that might be deployed to improve performance.

REFERENCES

Baddeley, A. (2003). Working memory: Looking back and looking forward. *Nature Reviews Neuroscience, 4*(10), 829–839.

Baron, J. (2000). *Thinking and deciding.* New York: Cambridge University Press.

Bartlett, F. C. (1932). *Remembering: An experimental and social study.* Cambridge, UK: Cambridge University.

Berbaum, K. S., Franken Jr., E. A., Dorfman, D. D., Rooholamini, S. A., Kathol, M. H., Barloon, T. J. et al. (1990). Satisfaction of search in diagnostic radiology. *Investigative Radiology, 25*(2), 133–140.

Bjork, R. A. & Bjork, E. L. (1992). A new theory of disuse and an old theory of stimulus fluctuation. *From learning processes to cognitive processes: Essays in honor of William K. Estes,* Vol. 2, pp. 35–67.

Brandimonte, M. A., Einstein, G. O. & McDaniel, M. A. (2014). *Prospective memory: Theory and applications.* Oxford: Psychology Press.

Broadbent, D. (1958). *Perception and communication.* London: Pergamon Press.

Cannon-Bowers, J. A. & Salas, E. E. (1998). *Making decisions under stress: Implications for individual and team training.* Washington, DC: American Psychological Association.

Carayon, P., Hundt, A. S., Karsh, B. T., Gurses, A. P., Alvarado, C. J., Smith, M. et al. (2006). Work system design for patient safety: The SEIPS model. *Quality and Safety in Health Care, 15*(1), i50–i58.

Custers, E. J., Regehr, G. & Norman, G. R. (1996). Mental representations of medical diagnostic knowledge: A review. *Academic Medicine, 71*(10), S55–S61.

Dekker, S. (2014). *The field guide to understanding "human error."* Surrey, UK: Ashgate.

Ericsson, K. A. & Lehmann, A. C. (1996). Expert and exceptional performance: Evidence of maximal adaptation to task constraints. *Annual Review of Psychology, 47*(1), 273–305.

Gardner, H. (1985). *The mind's new science: A history of the cognitive revolution.* New York: Basic Books.

Gawande, A. (2010). *The checklist manifesto: How to get things right,* Vol. 200. New York: Metropolitan Books.

Gigerenzer, G. & Goldstein, D. G. (1996). Mind as computer: Birth of a metaphor. *Creativity Research Journal, 9*(2–3), 131–144.

Holden, R. J., Carayon, P., Gurses, A. P., Hoonakker, P., Hundt, A. S., Ozok, A. A. et al. (2013). SEIPS 2.0: A human factors framework for studying and improving the work of healthcare professionals and patients. *Ergonomics, 56*(11), 1669–1686.

Kahneman, D. (2011). *Thinking, fast and slow.* New York: Macmillan.

Kahneman, D., Slovic, P. & Tversky, A. (1982). *Judgment under uncertainty: Heuristics and biases.* Cambridge, UK: Cambridge University Press.

Klein, G. (2008). Naturalistic decision making. *Human Factors, 50*(3), 456–460.

Kohn, L. T., Corrigan, J. M. & Donaldson, M. S. (2000). *To err is human: Building a safer health system.* Washington, DC: National Academies Press.

McDaniel, M. A. & Einstein, G. O. (2000). Strategic and automatic processes in prospective memory retrieval: A multiprocess framework. *Applied Cognitive Psychology, 14*(7), S127–S144.

Miller, G. A. (1956). The magical number seven, plus or minus two: Some limits on our capacity for processing information. *Psychological Review, 63*(2), 81.

Misasi, P., Lazzara, E. H. & Keebler, J. R. (2014). Understanding multi-team systems in emergency care, one case at a time. In Salas, E., Rico, R. & Shuffler, M. (Series Eds.), *Research on managing groups and teams: Vol. 17—Pushing the boundaries: Multi-team systems in research and practice.* Bradford, UK: Emerald.

Neisser, U. (2014). *Cognitive psychology: Classic edition.* New York: Psychology.

Pashler, H. (1994). Dual-task interference in simple tasks: Data and theory. *Psychological Bulletin, 116*(2), 220–244.

Poposki, E. M. & Oswald, F. L. (2010). The multitasking preference inventory: Toward an improved measure of individual differences in polychronicity. *Human Performance, 23*(3), 247–264.

Potchen, E. J. (2006). Measuring observer performance in chest radiology: Some experiences. *Journal of the American College of Radiology, 3*(6), 423–432.

Salas, E., Wilson, K. A., Lazzara, E. H., King, H. B., Augenstein, J. S., Robinson, D. W. et al. (2008). Simulation-based training for patient safety: 10 principles that matter. *Journal of Patient Safety, 4*(1), 3–8.

Schacter, D. L. & Tulving, E. (1994). *Memory systems.* Cambridge, MA: MIT Press.

Slovic, P. & Lichtenstein, S. (1971). Comparison of Bayesian and regression approaches to the study of information processing in judgment. *Organizational Behavior and Human Performance, 6*(6), 649–744.

Squire, L. R. & Zola, S. M. (1996). Structure and function of declarative and nondeclarative memory systems. *Proceedings of the National Academy of Sciences, 93*(24), 13515–13522.

Strauch, B. (2002). *Investigating human error: Incidents, accidents, and complex systems.* Burlington, VT: Ashgate.

Tsotsos, J. (1990). Analyzing vision at the complexity level. *Behavioral & Brain Sciences, 13*(3), 423–469.

Tuddenham, W. J. (1962). Visual search, image organization, and reader error in roentgen diagnosis: Studies of the psycho-physiology of roentgen image perception. *Radiology, 78,* 694–704.

Tversky, A. & Kahneman, D. (1974). Judgment under uncertainty: Heuristics and biases. *Science, 185*(4157), 1124–1131.

3 Situation Awareness, Sociotechnical Systems, and Automation in Emergency Medical Services

Theory and Measurement

David Schuster and Dan Nathan-Roberts

CONTENTS

INTRODUCTION

Decision-making, the selection of a choice among alternatives, is of critical importance in EMS. EMS are provided, under great time pressure, with many concurrent, high-stakes activities occurring simultaneously in a high-technology environment of practitioners (either providing care or operating an ambulance), dispatchers, and patients all affecting the outcome. In this environment, patient outcomes depend on timely and accurate human decision-making. Improving the decision-making of highly trained and high-performing professionals is a difficult challenge. In this chapter, we argue that decision-making in EMS can be improved through interventions targeting the immediate precursors of good decision-making. Situation awareness (SA), the degree to which one has actionable, goal-directed knowledge (Rousseau et al., 2004) of elements in the prehospital environment, provides a metric of precursors of effective decision-making in EMS.

Across domains, SA has been most effectively described as a predictor of quality decision-making in sociotechnical systems. It is distinct, but related to, preexisting knowledge, individual characteristics, workload, situational conditions, and human performance (Durso et al., 2006). That is, SA describes a state of holding relevant knowledge. It describes whether individuals have the knowledge needed in order to perform a task (Endsley, 1995). Consequently, SA provides a way to diagnose decision-making; in the moment, does an individual have the information needed to make the best decision?

Identifying deficiencies in SA and addressing them can lead to better decision-making, which, in turn, will improve patient outcomes. In the prehospital environment, however, individual decisions are not made in a vacuum. In order to positively impact the complex prehospital environment, human factor practitioners must consider how healthcare providers' behaviors interact within a complex, dynamic network of distributed team members, patients, technology, culture, and other factors. Thus, EMS is a *sociotechnical system*. Sociotechnical systems are frameworks that focus on a large system's overall performance through a high-level analysis of the individual components of the system and their interrelatedness as a means of understanding the system's performance as a whole (Holden et al., 2013). EMS can benefit from the structured tools of sociotechnical system analysis to better understand the impact of elements of the larger system on prehospital SA and decision-making.

In this chapter, we describe how EMS can be described as a sociotechnical system, some of the complexities and challenges inherent in sociotechnical systems, and the benefits of this perspective. Next, we summarize the state of the art in SA theory and practice as they apply to EMS. We will illustrate how SA fills a gap in our understanding of how individual human performance affects patient care. At the same time, we will consider the limitations of our current use of SA, both theoretically and in the field. This chapter will conclude with recommendations for the measurement of SA, the use of SA as a performance metric, and methods to predict the impact of interventions on SA.

EMS AS A SOCIOTECHNICAL SYSTEM

Sociotechnical systems are a way of understanding and improving systems at the large scale, macroergonomic level. Through the use of various high-level frameworks, such as macroergonomic analysis of structure, macroergonomic analysis design, and Systems Engineering Initiative for Patient Safety (SEIPS) 2.0 (Holden et al., 2013), sociotechnical systems provide a structured way to characterize a system or identify the components of the larger system with the most room for improvement. As it relates to EMS, sociotechnical system analysis is useful for looking holistically at the complexity of EMS across functional or organizational boundaries to identify gaps that might not be caught by using a more focused research lens. For example, the larger external cultural environment, such as friction between an administration and a union, may have impacts on the exchange of SA between system parts.

As an example, SEIPS, and its successor, SEIPS 2.0, which are arguably the most widely used sociotechnical models in healthcare, separate a large system into distinct subareas for analysis into work system, processes, and outcomes, with each area subdivided and interrelated to the others. Figure 3.1 shows the SEIPS 2.0 framework. The performance of EMS can be analyzed in terms of outcomes (patient, professional, and organizational) based on the processes used. To provide structure to the analysis of the EMS work system, the people, tools, and technology; organization; and internal environment (including environmental ergonomics, tasks, and external environment, including culture) are all studied as interacting components. Individual components and interactions can be studied using traditional human factors techniques, such as Rapid Upper Limb Assessment (McAtamney & Corlett, 1993) for tasks, or electronic health record usability (National Center for Human Factors in Healthcare, 2015) for technology.

This approach also provides analysis of the interactions among components. For example, poor physical ergonomics along with an internal culture that does not value worker safety may lead to undesirable patient outcomes by negatively impacting the physical processes of collaborative professional–patient work. Generally, the way in which components interact can be studied, as they provide barriers or facilitators to positive outcomes.

Healthcare is difficult to characterize because of the interconnectedness, varying levels of hierarchy, criticality of the temporal aspects, distributed nature of the teams, highly variable workload, and highly variable problems/procedures. Sociotechnical systems are often categorized by the tightness of coupling (how closely a change in one area affects a change in another area) and the level of criticality (how dangerous errors can be). For example, automotive manufacturing is a tightly coupled system, but the danger to human health of delaying the production line is rather low compared to healthcare. Conversely, sanitation departments play a critical role in our society, but they are not tightly coupled with the rest of the system. Unlike the automotive, sanitation, or energy domains, which also are heavy users of sociotechnical systems, healthcare is a very tightly coupled system with high criticality. However, this also means that healthcare has the most to gain from improvements found by reducing sociotechnical barriers. Work systems are often thought of as colocated and synchronized temporally. EMS teams can operate as distributed networks, with synchronous and asynchronous aspects. This difference from traditional inpatient care or traditional outpatient clinics makes EMS an even more tightly coupled, high-criticality system to study.

To properly analyze a system by using a sociotechnical framework, it is necessary to draw boundaries of analysis around the EMS or around the system under analysis. External to the EMS would be considered the external environment in a system model, and while it is proximal, it should be studied in sufficient detail only to provide information on the work system. In general, sociotechnical system analysis is not a panacea or recipe for analysis, but a holistic lens which can guide practitioners toward other human factors tool sets that can be used to examine and improve

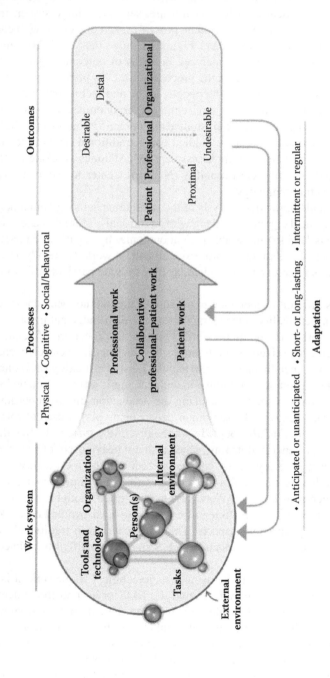

FIGURE 3.1   SEIPS 2.0 framework. (Reprinted from Holden, R.J. et al., *Ergonomics*, 56(11), 1669–1686, 2013. With permission.)

human decision-making. Next, we consider SA as one tool for sociotechnical system analysis.

SITUATION AWARENESS

SA was a concept described by fighter pilots before it was studied scientifically (Harwood et al., 1988). In general, SA describes goal-directed knowledge held by a decision maker, such as an emergency medical technician (EMT) (Rousseau et al., 2004). SA is distinguished from generalized knowledge in that it applies to the current task environment. SA is further distinguished from everything that could be known within an environment by the constraint that knowledge must support a goal. For example, a particular color of paint was used on the exterior of an ambulance, but such information is of no use in the care and transport of a patient. Thus, SA is defined by the goals of the individuals in the present situation. Measuring SA, then, first involves identification of all the possible goal-directed knowledge. The proportion of this knowledge held by an individual is SA.

ENDSLEY'S MODEL

Although there are competing models with various degrees of overlap, Endsley's (1995) model is dominant. It describes SA at three hierarchical but nonlinear levels (Endsley, 2015a). Level 1 is the *perception* of relevant elements in the environment. At level 1 SA, an individual attends to individual pieces of relevant information. Level 2 SA is the *comprehension* of the current situation; at this level, an individual connects pieces of information together in order to apply elements to the situational context. Level 3 SA is the *projection* of future status; it is knowledge of the state of elements in the future.

TEAM SA

The success of SA as a construct of applied cognition has come with challenges. While it has found particular success in aviation settings, a continued challenge has been the application of SA to teamwork. Because teamwork is more than the sum of its parts (Durso & Sethumadhavan, 2008) and teamwork in a sociotechnical system is affected by work system components, several approaches to measuring team cognition exist. In describing team-level SA, two approaches have received significant attention in the literature.

First is Endsley's concept of *team SA*. Team SA has been defined as "the degree to which every team member possesses the SA needed for his or her job" (Endsley, 1995, p. 39). It suggests a largely additive process in which individual practitioners are contributors to the team through a process of building individual SA. Shared SA is the overlap of the SA of individuals; it is where the SA of multiple individuals is equivalent. Shared SA is "the degree to which team members have the same SA on shared SA requirements" (Endsley & Jones, 2001, p. 48).

In this model, SA is held exclusively within human team members. This model extends beyond individual SA in the mechanisms by which individuals build SA.

Team process, identified by Salas et al. (1995), is a component of teamwork. Team process includes the communication and coordination behaviors that team members engage in as they perform taskwork. Endsley and Jones (2001) extended this work by identifying the requirements, devices, processes, and mechanisms of team SA. Communication, shared environments, and shared displays are the devices of team SA (Endsley & Jones, 2001). A limitation of this approach is that it can be difficult to model the complex impact of automation and other elements of the sociotechnical system if they do not cause an observable change in one individual's SA.

DISTRIBUTED SA

Stanton et al. (2006) argued for a different approach. Their concept of *distributed SA* treats SA as an artifact of the sociotechnical system. In this view, SA can be held in technological artifacts as well as by human team members. An example, such as that of a pulse oximeter, demonstrates the differences between these two perspectives. Is the status of the pulse oximeter sensor critical to one of the EMTs, or is it sufficient for the monitor to run without direct attention? Assuming that the oximeter provides an alert, is perfectly accurate, and is not being read, information held in the monitor would not be part of team SA. It would, however, be considered a part of distributed SA in that it is relevant information held by a technological agent (the meter). When the EMT must attend to an alert or read the meter, this information becomes part of that EMTs individual SA or, when told to another team member, part of shared SA. Under the distributed SA perspective, reading the value from the meter causes that information to be shared between the technological agent and the EMT. Endsley (2015a, p. 26) argued against a distributed approach to SA, suggesting that only automation that is a "cognizant and independent decision maker" could be considered to have SA. Since the technology used in ambulances is becoming ever more sophisticated and decisions in sociotechnical systems are so tightly coupled, this line is not a clear one. Distributed SA is attractive because it provides a way to describe the benefits of automating information processing. However, distributed SA measures the degree to which information is managed within the sociotechnical system.

We repeat an argument indicating that these perspectives may be partially reconciled by quantifying both overlapping and nonoverlapping information and allowing for agents to be diverse in their information processing (Cain & Schuster, 2014; Ososky et al., 2012). That is, both people and technology may contribute to SA, but cognition is a unique property of people. Individual SA describes the knowledge held by one individual, with some of this information shared by others. Individual SA that overlaps is shared; individual SA held by one agent, human or technological, but relevant to others is complementary SA.

While it is important for researchers and practitioners to understand the theoretical complexity surrounding SA, much of the theoretical debate is of limited relevance to the practitioner. What is critical, regardless of the theoretical approach, is that SA reflects goal-directed knowledge. The proper definition of goals is critical. If goals include good decision-making on the part of an individual human, then an individual's SA must be defined based on the information needed for individual decision-making. However, we have made a case for a sociotechnical system approach to SA in EMS regardless of the theoretical approach.

ISSUES IN EMS DECISION-MAKING AND RECOMMENDATIONS

ISSUE: SA DEFINITION CONFUSION AND MULTIPLE MEASUREMENT TECHNIQUES

Both SA and human performance have been defined in many ways, leading to confusion about the use of SA. Parasuraman et al. (2008) argued that SA is distinct from generalized knowledge, task performance, and the quality of decision-making. They suggested that SA is distinctly useful in atypical circumstances. In post hoc investigations of accidents in domains such as aviation and driving, SA has been frequently found to be a contributing factor (Jones & Endsley, 1996; Durso & Sethumadhavan, 2008). Some evidence for this claim is found in empirical work examining differences between experts and novice pilots. Individual SA is a construct with well-established predictive validity (Durso et al., 2006) and a widely cited model, but it is difficult to model in general terms without context (Wickens, 2015). In EMS, Endsley's model can be used to diagnose the cognitive performance of practitioners within the sociotechnical system despite debates about the meaning and validity of subcomponents of Endsley's model (e.g., Flach, 1995; Hoffman, 2015). Therefore, practitioners should weigh the relatively small effort of conducting goal-directed task analysis against the utility of capturing goal-relevant knowledge, since SA predicts a variety of diverse performance outcomes. While it is tempting to identify everything that could possibly be known in an environment and reward those who hold the greatest amount of information during task performance, SA is an appropriate metric only when it targets specific goals rather than all possible outcomes. Appropriate measurement SA should focus exclusively on the knowledge needed to immediately achieve such goals. The goal-oriented nature of SA distinguishes it from trivia.

There are a variety of measurement techniques available. The situation awareness global assessment technique (SAGAT) is an example of a measurement technique in which a task or simulation is paused and then individuals are prompted to answer questions that sample SA (Endsley, 1988, 2000). SPAM, the situation present assessment method (Durso & Dattel, 2004), is a variation of this technique in which questions are made available randomly with a signal. The individual pushes a button to view and respond to the question as soon as they are able, but the task environment continues uninterrupted. Questions could be offered on a tablet or other mobile device. Neither SAGAT nor SPAM measurement would be suitable in an operational environment, but both are well suited to simulation and training environments.

SPAM offers a way to simultaneously measure workload; question response latency can be used as a proxy measure of workload. When operators are under high demand, they may take longer to respond to the probe questions (Durso et al., 2006). Quantifying the time to question can provide a supplementary workload measure without requiring a separate questionnaire, although this method may be less sensitive than other assessments of workload (Pierce et al., 2008).

SA measures that are not domain specific also exist, but they suffer from validity issues. The most common of these, the situation awareness rating technique (SART) (Taylor, 1990) is a self-report questionnaire that asks an individual to report their levels of awareness along several dimensions: supply of attentional resources, demand on attentional resources, and understanding of the present situation. It is administered after task performance. If an individual lacks SA but does not know what they do not know, they may rate their SA highly. Further, high-performing individuals may be better able to think about their own SA losses and could rate themselves lower than our oblivious individual. In this way, SART is more of a metacognitive measure than an SA measure. Empirical data have provided support to these theoretical criticisms (Endsley, 1988). Although the SART is the least costly measure to implement, the validity issues make it less useful for operational use in EMS.

Recommendation: Unless such measures already exist, use goal-directed task analysis to facilitate the use of domain-specific measures of SA. Goal-directed task analysis is a method for describing the knowledge requirements in a situation (Endsley & Jones, 2012). This method requires access to high-performing individuals to identify their goals and subgoals. The result of this analysis can be represented as a hierarchical list of information requirements that can be classified at the three levels of SA. This technique can also identify shared SA requirements (Bolstad et al., 2002). It is important that a continuum of expertise is represented in goal-directed task analysis, as experience brings more efficient use of working memory and better strategies (Endsley, 2015b).

Goal-directed task analysis is a prerequisite for defining ideal SA in a new environment. While goal-directed task analysis results in a hierarchical list of information without description of how that information is obtained, it is inherently bound to an operational context. Consequently, SA measures may need revision as prehospital teams vary in their composition or as technology affects what knowledge must be held by team members in the prehospital environment.

Issue: Individual SA Provides an Incomplete Picture

Individual SA will likely predict the quality of patient care, but it is not the whole story. To properly capture SA, practitioners and researchers should take care to broadly consider SA in the sociotechnical system. Because SA is inherently tied to tasks, the boundaries of the sociotechnical system (i.e., the situation) must be carefully delineated. In EMS, the activities within the ambulance are at one point along the care continuum but exclude events related to dispatch and transfer to the emergency department. Within the situation, individual decision makers should be identified. Individual decision makers will include EMS practitioners, but they should also include other human team members such as physicians and dispatchers. Additionally, the contribution of any technology capable of decision-making should also be included.

Recommendation: Capture team process to contextualize individual SA metrics. Despite theoretical debates, application of Endsley's (1995) model is likely to be a useful approach. However, an analysis that identifies gaps in individual SA can be misleading without identifying deficiencies in team process behaviors, including how information is communicated (Cooke et al., 2007). As members of a team, loss of SA by an individual is unlikely to be an isolated problem. Both the antecedents of SA loss and the impacts of SA loss are intertwined with other elements of the socio-technical system. Consequently, the goal should not be to single out low-performing individuals but to understand how individual performance is hindered or facilitated by elements of the sociotechnical system.

At one end of a spectrum, the simplest approach to measurement is quantitative assessment of the SA provided by human team members and technological decision aids. Individual SA measurement can be augmented by quantitative or qualitative analysis of team process behaviors. At the other end of this spectrum is a distributed approach to SA measurement, which focuses to the greatest degree on the interactions among elements in the sociotechnical system. A distributed approach to SA measurement may be possible with a well-defined situation model and a thorough understanding of the contribution of technology to individual SA. At present, empirical work is still needed to support a comprehensive understanding of EMS as a sociotechnical system.

Issue: Automation in the Ambulance Can Have Unanticipated Impacts on SA

Since its inception, research on SA has been applied to human–technology interaction, starting with the aviation cockpit. Other domains in which SA has been effectively applied frequently include automated tools. EMS is no exception with increasingly sophisticated medical equipment becoming a part of prehospital care. An example is the increasing computerization of medical equipment and the use of diagnostic aids, such as a recent National Institutes of Health Stroke Scale Assessment administered on a tablet (Padrick et al., 2015).

In other domains, empirical results have suggested, on the surface, that automated tools reduce workload and increase performance. Subsequent research has suggested that SA moderates this relationship (Wickens et al., 2010). However, high levels of automation can lead to a loss of SA (Endsley & Kiris, 1995). Consequently, SA is useful in diagnosing the impact of automation on decision-making performance. Several recurring issues in human–automation interaction are common in sociotechnical systems across domains. Fortunately, an understanding of the potential issues introduced by automation can inform automation design to minimize their impact.

First, automation may increase workload by placing additional demands on the practitioner. In doing so, SA is likely to be reduced (Vortac et al., 1993). Since critical decisions are regularly made under conditions of high mental and physical workload, it is important that new tools do not place an additional burden on already taxed resources. Another issue is the out-of-the-loop performance problem (Endsley & Kiris, 1995), which characterizes a situation of high-performing, but imperfect automation. Automation can be imperfect by occasionally failing outright, such as in the case of a malfunctioning sensor that reports either no value or an incorrect value. Automation can also be imperfect in that it fails to be usable in complex, high-workload situations. In either case of imperfect automation, human operators find themselves largely able to rely on the automation, except during rare occurrences. In these rare occurrences, humans are called "back into the loop" and must take over physical or cognitive tasks typically performed by the automation. People are particularly bad at a rapid, unanticipated shift to manual control. The result is a loss of SA.

Recommendations: Avoid EMS automation that overpromises and underdelivers. Provide training to users of automation on the true capabilities and limitations of automation to encourage appropriate trust. Train for mitigation procedures under automation failure.

Notably, such a loss of SA might not be observed in the case of poorly functioning automation. When working with automation, people base trust attributions on their perceptions of automation performance (Johnson et al., 2009; Muir & Moray, 1996; Oleson et al., 2011, p. 176). Automation which is known to be unreliable may be relied upon less (Yeh & Wickens, 2001). Although this may seem to mitigate the out-of-the-loop performance problem, it may be more accurately described as a "stuck-in-the-loop" performance problem. Such automation is either appropriately disregarded, in which case it provides no benefit, or inappropriately underutilized, a condition called disuse (Parasuraman & Riley, 1997). When human operators of otherwise useful automation elect not to use it, the problem is a lack of trust in automation (Lee & See, 2004; Parasuraman & Riley, 1997). Trust in automation is "the attitude that an agent will help achieve an individual's goals in a situation characterized by uncertainty and vulnerability" (Lee & See, 2004, p. 54). This state of affairs can become evident during goal-directed task analysis. Experts, as part of task analysis, may identify technology that is rarely used as intended due to its perception as being ineffective. When trust is appropriately calibrated, it is adaptive. Trust allows people to appropriately use automated tools.

To encourage appropriate calibrated trust, it is important that the performance of automation in the prehospital environment matches practitioner perceptions of its performance. All other factors being equal, trust in automation will be higher with higher levels of automation reliability. Training that overstates the capabilities of automation may have the effect of reducing use of the automation. For many forms of automation, perfect reliability is not possible. If automation can require unanticipated human intervention, especially during periods of time pressure and high workload, practitioners should be trained for these scenarios to minimize the effects of the out-of-the-loop performance problem.

CONCLUSIONS

Our aim in this chapter was to connect EMS to decades of research on SA. As a sociotechnical system, measurement of SA in EMS can be more complex than in other domains. However, several theoretical perspectives and approaches are available to practitioners using SA as a performance metric. Whether using a team SA or distributed SA approach, measurement of SA can capture the quality of teamwork and the emergent cognition of the sociotechnical system. However, this approach requires the greatest understanding of how the diverse elements in the sociotechnical system (including team members, patients, and technologies) interact. Alternatively, a less nuanced approach to SA can incorporate individual SA measures along with considerations of other components of the sociotechnical system, such as human–automation interaction. Although each provides only one piece, together they can diagnose the precursors of performance in EMS.

REFERENCES

Bolstad, C. A., Riley, J. M., Jones, D. G. & Endsley, M. R. (2002). Using goal directed task analysis with Army brigade officer teams. *Proceedings of the Human Factors and Ergonomics Society Annual Meeting*, 472–476. Santa Monica, CA: Human Factors and Ergonomics Society.

Cain, A. A. & Schuster, D. (2014). Measurement of situation awareness among diverse agents in cyber security. *Proceedings of the IEEE International Inter-disciplinary Conference on Cognitive Methods in Situation Awareness and Decision Support (CogSIMA)*, 124–129. San Antonio, TX: Institute of Electrical and Electronics Engineers.

Cooke, N. J., Gorman, J. & Winner, J. (2007). Team cognition. In F. Durso, R. Nickerson, S. Dumais, S. Lewandowsky & T. Perfect (Eds.), *Handbook of applied cognition* (2nd ed., pp. 239–268). Hoboken, NJ: Wiley.

Durso, F. T., Bleckley, M. K. & Dattel, A. R. (2006). Does situation awareness add to the validity of cognitive tests? *Human Factors*, *48*(4), 721–733.

Durso, F. T. & Dattel, A. R. (2004). SPAM: The real-time assessment of SA. In S. Banbury & S. Tremblay (Eds.), *A cognitive approach to situation awareness: Theory and application* (pp. 137–154). Aldershot, UK: Ashgate.

Durso, F. T. & Sethumadhavan, A. (2008). Situation awareness: Understanding dynamic environments. *Human Factors*, *50*(3), 442–448.

Endsley, M. R. (1988). Design and evaluation for situation awareness enhancement. *Proceedings of the Human Factors and Ergonomics Society Annual Meeting*, 97–101. Santa Monica, CA: Human Factors and Ergonomics Society.

Endsley, M. R. (1995). Toward a theory of situation awareness in dynamic systems. *Human Factors*, *37*(1), 32–64.

Endsley, M. R. (2000). Direct measurement of situation awareness: Validity and use of SAGAT. In M. R. Endsley & D. J. Garland (Eds.), *Situation awareness analysis and measurement* (pp. 147–173). Mahwah, NJ: Lawrence Erlbaum Associates.

Endsley, M. R. (2015a). Situation awareness misconceptions and misunderstandings. *Journal of Cognitive Engineering and Decision Making*, *9*(1), 4–32.

Endsley, M. R. (2015b). Final reflections: Situation awareness models and measures. *Journal of Cognitive Engineering and Decision Making*, *9*(1), 101–111.

Endsley, M. R. & Jones, D. G. (2012). *Designing for situation awareness: An approach to human-centered design* (2nd ed.). London: Taylor & Francis.

Endsley, M. R. & Jones, W. M. (2001). A model of inter- and intrateam situation awareness: Implications for design, training and measurement. In M. McNeese, E. Salas & M. Endsley (Eds.), *New trends in cooperative activities: Understanding system dynamics in complex environments* (pp. 46–67). Santa Monica, CA: Human Factors and Ergonomics Society.

Endsley, M. R. & Kiris, E. O. (1995). Toward a theory of situation awareness in dynamic systems. *Human Factors*, *37*(2), 381–394.

Flach, J. M. (1995). Situation awareness: Proceed with caution. *Human Factors*, *37*, 149–157.

Harwood, K., Barnett, B. & Wickens, C. (1988). Situational awareness: A conceptual and methodological framework. Paper presented at the meeting of the *11th Symposium of Psychology* in the Department of Defense, Colorado Springs, CO.

Hoffman, R. (2015). Origins of situation awareness: Cautionary tales from the history of concepts of attention. *Journal of Cognitive Engineering and Decision Making*, *9*(1), 73–83.

Holden, R. J., Carayon, P., Gurses, A. P., Hoonakker, P., Hundt, A. S., Ozok, A. A. & Rivera-Rodriguez, A. J. (2013). SEIPS 2.0: A human factors framework for studying and improving the work of healthcare professionals and patients. *Ergonomics, 56*(11), 1669–1686.

Johnson, R. C., Saboe, K. N., Prewett, M. S., Coovert, M. D. & Elliott, L. R. (2009). Autonomy and automation reliability in human–robot interaction: A qualitative review. *Proceedings of the Human Factors and Ergonomics Society 53rd Annual Meeting, 53*, 1398–1402. Santa Monica, CA: Human Factors and Ergonomics Society.

Jones, D. G. & Endsley, M. R. (1996). Sources of situation awareness errors in aviation. *Aviation, Space, and Environmental Medicine, 67*(6), 507–512.

Lee, J. D. & See, K. A. (2004). Trust in automation and technology: Designing for appropriate reliance. *Human Factors, 46*(1), 50–80.

McAtamney, L. & Corlett, E. N. (1993). RULA: A survey method for the investigation of work-related upper limb disorders. *Applied Ergonomics, 24*(2), 91–99.

Muir, B. M. & Moray, N. (1996). Experimental studies of trust and human intervention in a process control simulation. *Ergonomics, 39*(3), 429–460.

National Center for Human Factors in Healthcare. (2015). EHR User-Centered Design Evaluation Framework. Retrieved from http://healthitusability.org.

Oleson, K. E., Billings, D. R., Kocsis, V., Chen, J. Y. C. & Hancock, P. A. (2011). Antecedents of trust in human–robot collaborations. *IEEE International Multi-disciplinary Conference on Cognitive Methods in Situation Awareness and Decision Support (CogSIMA)*, 175–178. Miami, FL: Institute of Electrical and Electronics Engineers.

Ososky, S., Schuster, D., Jentsch, F., Fiore, S., Shumaker, R., Lebiere, C. et al. (2012). *The importance of shared mental models and shared situation awareness for transforming robots from tools to teammates.* Baltimore, MD: International Society for Optics and Photonics.

Padrick, M. M., Chapman Smith, S. N., McMurry, T. L., Mehndiratta, P., Chee, C. Y., Gunnell, B. S. et al. (2015). NIH stroke scale assessment via iPad-based mobile telestroke during ambulance transport is feasible: Pilot data from the Improving Treatment with Rapid Evaluation of Acute Stroke via mobile telemedicine (iTREAT) study. *Stroke, 46*(Suppl 1), A90.

Parasuraman, R. & Riley, V. (1997). Humans and automation: Use, misuse, disuse, abuse. *Human Factors, 39*(2), 230–253.

Parasuraman, R., Sheridan, T. B. & Wickens, C. D. (2008). Situation awareness, mental workload, and trust in automation: Viable, empirically supported cognitive engineering constructs. *Journal of Cognitive Engineering and Decision Making, 2*(2), 140–160.

Pierce, R. S., Vu, K.-P. L., Nguyen, J. & Strybel, T. Z. (2008). The relationship between SPAM, workload, and task performance on a simulated ATC task. *Proceedings of the Human Factors and Ergonomics Society Annual Meeting*, 34–38. Santa Monica, CA: Human Factors and Ergonomics Society.

Rousseau, R., Tremblay, S. & Breton, R. (2004). Defining and modeling situation awareness: A critical review. In S. Banbury & S. Tremblay (Eds.), *A cognitive approach to situation awareness: Theory and Application* (pp. 3–21). Burlington, VT: Ashgate.

Salas, E., Prince, C., Baker, D. P. & Shrestha, L. (1995). Situation awareness in team performance: Implications for measurement and training. *Human Factors, 37*(1), 123–136.

Stanton, N. A., Stewart, R., Harris, D., Houghton, R. J., Baber, C., Mcmaster, R. et al. (2006). Distributed situation awareness in dynamic systems: Theoretical development and application of an ergonomics methodology. *Ergonomics, 49*(12–13), 1288–1311.

Taylor, R. M. (1990). Situational awareness rating technique (SART): The development of a tool for aircrew systems design. (AGARD-CP-478) Neuilly Sur Seine, France: NATO-AGARD.

Vortac, O. U., Edwards, M. B., Fuller, D. K. & Manning, C. A. (1993). Automation and cognition in air traffic control: An empirical investigation. *Applied Cognitive Psychology*, *7*(7), 631–651.

Wickens, C. D. (2015). Situation awareness: Review of Mica Endsley's 1995 articles on situation awareness theory and measurement. *Human Factors*, *50*(3), 397–403.

Wickens, C. D., Li, H., Sebok, A. & Sarter, N. B. (2010). Stages and levels of automation: An integrated meta-analysis. *Proceedings of the Human Factors and Ergonomics Society 54th Annual Meeting*, *54*, 389–393. Santa Monica, CA: Human Factors and Ergonomics Society.

Yeh, M. & Wickens, C. D. (2001). Display signaling in augmented reality: Effects of cue reliability and image realism on attention allocation and trust calibration. *Human Factors*, *43*(3), 355–365.

4 Naturalistic Decision-Making in Emergency Medical Services

Michael A. Rosen, Ian Coffman, Aaron Dietz,
P. Daniel Patterson, and Julius Cuong-Pham

CONTENTS

INTRODUCTION

EMS workers navigate an environment filled with uncertainty, high stakes, time pressure, complexity, and, at times, conditions that pose risk to their personal safety all while managing complex patient care tasks. They make life and death decisions under these extreme conditions. An EMS worker must diagnose injuries and diseases, assess risks, prioritize treatment options, and manage the social environment. The process of decision-making is central to all of these tasks. Therefore, a detailed understanding of how expert EMS workers make decisions can be invaluable for designing systems to support their work as well as developing new expert EMS workers as efficiently as possible.

EMS clearly represents a unique and demanding work domain. However, parallels can be drawn to other complex, high-risk, and time-pressured contexts. Military, aviation, firefighting, policing, power generation, financial services, and intelligence analysis domains are some key industries and professions that share some of the EMS characteristics and constraints. Naturalistic decision-making (NDM) is an applied research tradition seeking to understand and improve decision-making processes in these types of situations. This approach is rooted in the science of decision-making and human expertise and values the importance of context for truly understanding and improving decision-making in professional work. In this chapter, we use NDM as a framework to review and integrate existing research on EMS worker decision-making. Specifically, we address three key goals. First, we present a brief review of the NDM literature, focusing on the performance mechanisms that underlie expert performance and development. Second, we review the current literature related to EMS decision-making to provide recommendations for work system design. Third, we discuss implications for training and learning interventions to accelerate the development of expert EMS decision makers.

FUNDAMENTAL CONCEPTS, DEFINITIONS, AND METHODOLOGIES OF NDM

The NDM tradition emerged in the 1980s from the practical need to understand how people make decisions outside of artificial laboratory experiments which dominated the decision-making research at the time. The stark contrasts between NDM and traditional decision-making paradigms have been well documented (see Cannon-Bowers et al., 1996; Lipshitz et al., 2001; Orasanu, 2005; Klein, 2008; Rosen et al., 2008; Grossman et al., 2014). The classical decision-making perspective theorized decision-making as a process of selecting the "best or optimal choice" from an exhaustive assemblage of alternatives. The decision maker would evaluate the relative merits of all alternatives in relation to prescribed criteria, or standards, and decisions made based on which option yields the most promising outcome. This approach assumed that the decision maker was not constrained in terms of information and time; that is, that all relevant information was known at the time of the decision and that a thorough analysis of all options was possible. In a slight divergence from this view, rational choice models recognized that decision makers may stray from optimal decision strategies resulting from the application of cognitive heuristics—or general rules applied to reduce the complexity of decision-making. Rationalists also postulated that the decision-making process consisted of concurrent option selection and regarded deviations from the most favorable strategy as errors. The belief that normative models shape ideal decision-making is fundamental to both of these perspectives.

These models, however, proved impractical and ineffective when applied to the task of understanding how decision-making happens in the types of situations faced by many professionals (Klein, 2008). Consider the prehospital care environment, which may involve time pressure, ambiguity or incomplete information surrounding the condition of the patient(s) when arriving on the scene, varying degrees of patient complexity from case to case, evolving circumstances, and the need to interact with other clinicians on- and off-site (Gunnarsson & Stomberg, 2009). Under such conditions, it would be unrealistic to evaluate a fixed set of alternatives based on probabilistic consequences associated with each choice as conceived by traditional decision-making models. In fact, there may not even be a paragon for how to proceed but simply a good choice for how to respond to a situation that is derived from experience, recognizing critical cues, and acting accordingly. The constraints and complexity faced by professionals in these situations and the desire to provide tools to support their work spawned the NDM movement.

At its core, NDM is concerned with how experts make decisions and how attributes of the environment shape the decisions that are made (Zsambok, 1997). Central to this perspective is that (1) uncertainty, time pressure, and high stakes are intrinsic to real-world decision-making; (2) the expertise of the decision maker is paramount; and (3) experts do not simply select the "best choice" for how to proceed but match their decisions with what is appropriate for a given situation (Lipshitz et al., 2001). Naturalistic studies seek to understand decision-making in situ because contextual characteristics often dictate the responses that are possible and effective. As described by Orasanu and Connolly (1993), NDM settings are characterized by uncertainty and dynamic environments, shifting and ill-defined or competing goals, action/feedback loops, time constraints, high stakes, multiple players, and organizational goals and norms (Rosen et al., 2008, p. 213). While each of these factors manifests in many settings, not all of them need to be present in a situation for it to be considered a naturalistic context. The prehospital environment, for instance, may feature different factors from case to case.

NDM also emphasizes expertise because it considers the cognitive processes underlying decision-making. Rather than trying to predict a specific option that should be implemented, NDM models seek to understand what task and environmental features that experts look for, how meaning is ascribed to them, and the strategies that are employed when making decisions (Lipshitz et al., 2001). According to Klein's recognition-primed decision (RPD) model, expert decision-making takes place through pattern recognition and mental simulations (Klein, 1993a). First, the decision maker matches situational cues with prior experiences to determine whether previous decision strategies are applicable. This occurs rapidly and establishes expectations, objectives, and possible responses for how to proceed (Klein, 2008). Next, the decision maker engages in a mental simulation to evaluate whether the strategy will work. This results in adopting the strategy, adapting the strategy, or rejecting the strategy all together. If the strategy is rejected, the decision maker seeks additional cues for how best to proceed.

Decision-making as explicated by the RPD model is a combination of intuitive and deliberative information processing; pattern matching involves making unconscious, immediate appraisals of the situation, while mental simulation is slow, cognizant, and methodical (Klein, 2008). While both processes are crucial for making effective decisions, intuition expeditiously facilitates the identification and integration of critical task features. A purely deliberative strategy would be appropriate for simple tasks with explicit criteria for success/failure, although it would place a heavy cognitive burden when decision makers are confronted with the ill-structured problems, dynamic circumstances, and time pressures characteristic of NDM environments.

Similarly, intuition is not a silver bullet for quickly reaching an effective decision; capitalizing on intuition depends on extensive experience within a given work domain (Salas, Rosen & DiazGranados, 2010). Salas, Rosen, and DiazGranados (2010) introduced the concept of expertise-based intuition to describe the effective use of intuition in decision-making. Expertise-based intuition is defined as "intuition rooted in extensive experience within a specific domain" (p. 941). This extensive experience affords the development of a rich knowledge base from which to guide decision-making. Experts draw from a larger and more sophisticated repository of situational patterns stored in their long-term memory compared to novices (Gobet & Simon, 1996). When confronted with a problem, representations are linked with these structures, providing experts with more efficient access to their knowledge (Ericsson & Charness, 1994).

NDM Methodology

Practical applications of NDM research may involve designing improved systems or training decision makers. Before this can take place, methods to model the cognitive (e.g., situation assessment, pattern recognition, automaticity) and domain-relevant (e.g., difficulty, frequency, time pressure) factors underlying expert decision-making must be employed. Cooke (1999) outlined a collection of elicitation techniques to this end, including observations, interviews, process tracing, and conceptual methods. Observations are particularly useful when first getting immersed into a domain and can help guide subsequent knowledge elicitation approaches that are more explicit. Observations can take many forms (e.g., passive observation, participatory observation, video analysis) and vary in focus (e.g., examination of specific role or event). An advantage to observation is studying decision-making in context, but simulations may provide a substitute when task conditions preclude naturalistic observation.

Interviews with domain experts also constitute a rich source of information. Unstructured interviews involve a natural, unrestricted discussion, while structured interviews follow an explicit procedure (i.e., same questions are asked to all participants in the same order). Semistructured interviews fall somewhere within this spectrum, allowing greater flexibility to discuss certain topics in more or less detail. The critical decision method (CDM) is a widely applied interview strategy in NDM research (see Klein, Calderwood & MacGregor, 1989, for a full description). Based on Flanagan's (1954) critical incident technique, CDM involves constructing a timeline around an event. Probes are embedded within the interview sequence to tap more specific information that influenced key decision points during the event, such as the decision maker's experience (e.g., what aspects of your training helped you make that decision?), analogues (e.g., did that remind you of a similar experience?), and goals (e.g., what was your objective?).

Process tracing involves capturing the sequence of behavioral events associated with completing a task and is generally applied to obtain procedural information. A popular process tracing approach is the think-aloud method, in which participants give verbal reports of their behaviors as they are executing a task. Last, conceptual methods seek to characterize how concepts within a given domain are related. This entails (1) identifying concepts (e.g., through observations and interviews),

(2) determining the extent to which experts feel that they are related (e.g., relatedness ratings), (3) scaling down and making a visual representation of concept data (e.g., multidimensional scaling, pathfinder network scaling, soliciting experts to draw a graphical representations), and (4) evaluating the reduced conceptual model.

IMPLICATIONS FOR SYSTEM DESIGN

In this section, we describe how such an understanding can be applied to the design of safe work systems. We define a *work system* as anything in the EMS worker's environment that plays a substantive role in work-related performance, including but not limited to tools and devices, interactions with other team members, established protocols, etc. Space limitations prohibit a complete discussion of work systems and the potential impact of NDM-based design principles, so we instead outline a general, extensible framework for the integration of theory-driven approaches to decision making and safe system design in the EMS setting.

We start by briefly describing five core performance mechanisms of expert decision-making as identified in the NDM literature: situation assessment and problem representation, pattern recognition, sensemaking, mental simulation, and automaticity. We then provide a worked-out example of how these principles viewed in the context of an NDM-based design framework can be used to inform the design of safe work systems.

SITUATION ASSESSMENT AND PROBLEM REPRESENTATION

Situation assessment and problem representation refer to the ability of expert decision makers to quickly and effectively extract the crucial features of a situation or problem in order to generate and implement a course of action. Under an NDM-based interpretation of this process, features of the decision maker's expertise such as prior experience and domain-specific knowledge act as a set of a priori constraints on the space of possible courses of action. In the extreme case, the decision maker's situation assessment and representation abilities may be so well tuned that only a single, optimal course of action results, following as a logical consequence of their familiarity with the situation (Klein, 1993b). This stands in contrast to decision-making approaches which posit that all possible courses of action are simultaneously considered and pared down relative to the demands of the situation at hand. Randel et al. (1996) found support for this contrast by comparing the behavior of novice and expert electronic warfare technicians. They found that while novices tend to approach a situation by immediately deciding on a course of action, expert technicians engage in situation assessment and problem representation from which a course of action naturally follows.

PATTERN RECOGNITION

Expert decision makers engage in pattern recognition by mentally organizing features of an environment or task into a set of meaningful, general units. This is necessary because humans, regardless of their expertise, possess limited working memory abilities and are thus unable to consider all aspects of a situation at an extreme level of detail. A core mechanism by which this detail is reduced to meaningful units is known as *chunking* (Miller, 1956). For example, Chase and Simon (1973) showed that expert chess players reduce complex game situations into chunks consisting of groups of pieces organized by game-based constraints (e.g., possible moves). Thus, experts are able to rein in the complexity of a situation by mentally organizing it into a set of smaller, manageable patterns, while novices may lack this ability. Similar to the mechanism of situation assessment and problem representation discussed earlier, experts' pattern recognition processes are automatic and do not require deliberate action (Drews & Kramer, 2012). Expert decision makers are ultimately able to compare these patterns to those already familiar from experience to aid in selecting the best course of action.

SENSEMAKING

Until now, we have been discussing the mechanisms by which experts are able to make decisions in the context of their prior knowledge or experience with a situation or problem. But expert decision makers are also able to quickly and effectively react to novel or out-of-the-ordinary situations through a process called sensemaking. Sensemaking is the performance mechanism by which experts are able to develop the "big picture," including identifying the problem, forming possible solutions, and making inferences about potential outcomes (Weick, 1993). Sensemaking is utilized when features of the situation at hand fail to align with the expert's prior experiences or expectations (Klein et al., 2007).

MENTAL SIMULATION

Expert decision makers internally simulate possible courses of action to arrive at a set of possible outcomes from which the best is selected. Mental simulation relies on *mental models*, which are cognitive representations of the external world (Craik, 1943). Mental models are utilized as the basis of mental simulation, providing an internal testing ground in which decision makers can mentally implement possible courses of action and predict their outcomes. These models are not necessarily high in fidelity and may feature characteristics of the environment or task that are of particular relevance to the expert's task, including specialized knowledge (Webber et al., 2000). Klein and Crandall (1995) found that the majority of mental simulation instances that they considered made reference to visual imagery, which suggests that experts may utilize mental simulation particularly when the task or problem at hand deals with primarily visual information, such as determining the injection site for an intravenous (IV) needle. Like the other mechanisms discussed earlier, mental simulation is an ability that experts have honed on the basis of their existing knowledge and prior experiences, meaning that novices may rely less on it or perform these simulations with less success.

AUTOMATICITY

The more practice and experience one gains with performing a particular task, the easier it is for that task to be performed with less deliberate effort. Like the process of pattern recognition discussed previously, automation is another strategy for dealing with limited cognitive resources by allowing the decision maker to expend more effort on aspects of the task at hand that might be less well practiced or familiar. An additional benefit of automation is that the automated tasks may be performed more quickly and accurately (Moors & De Houwer, 2006). The kinds of tasks automated by experts tend to include those that are the most basic or frequently performed in a given domain (Lesgold et al., 1988).

NDM-BASED DESIGN IN PRACTICE: THE CASE OF DRUG ADMINISTRATION

With the core performance mechanisms established, we turn our attention to their application in a real-life scenario. The administration of medication in EMS settings is a complex task that must be performed quickly and accurately in the presence of additional stressors such as time constraints and limited diagnostic information. It is also an area prone to a high incidence of errors: Hubble et al. (2000) report that for the paramedics that they tested, the mean success rate of dosage and delivery calculations was only 51.4%. These results were obtained from a pencil-and-paper-based examination, so performance would likely degrade further in more realistic, stressful situations. Schwartz et al. (2006) discuss a number of additional issues contributing to drug administration errors in EMS settings, including drug identification issues stemming from illegible and inconsistent labeling and confusion caused by similar-looking medications. The issue addressed in this section is how an NDM-based approach to work system design could potentially reduce errors in drug administration.

While Hubble et al.'s (2000) findings on the poor performance of EMS workers on drug dosage calculations present a clear illustration of a problem, their results offer little insight into the cognitive mechanisms underlying paramedics' approach to drug administration situations. This issue is partially illuminated by Feufel et al.'s (2009) finding that EMS workers tend to employ a set of broad, relatively informal heuristics for dealing with these situations (e.g., smaller doses for children and larger doses for adults) rather than undertaking precise calculations on the basis of weight, age, etc. An NDM-based solution to this problem should ideally acknowledge and engage this kind of heuristic and others like it. We now address how the particular issues discussed in the preceding paragraph might serve as the basis for the design of a work system aimed at limiting the occurrence of drug administration errors.

A number of the performance mechanisms discussed earlier are implicated in the task of drug administration. The paramedics studied by Feufel et al. (2009) engage in situation assessment and problem representation and pattern recognition by comparing general features of the patient (e.g., child versus adult) with their prior experiences relating to the appropriate drug dosages required by patients displaying those features. A general means by which work systems can integrate such processes is by translating them as well as possible into elements of the external environment. For example, medications could be physically or spatially organized in a manner consistent with paramedics' intuitions about proper dosages relative to a patient's general features, potentially reducing the number of calculations required. For example, the Broselow tape is a widely adopted method of a visually oriented system designed to aid in assessment of drug dosages appropriate for pediatric patients and organization of medication carts to expedite administration of drugs (Agarwal et al., 2005). This approach helps to structure the work environment to reduce unnecessary variation in the environment and to embed knowledge into the job tools.

Injecting IV medication, for example, is an example of a well-trained procedural skill possessed by paramedics and, thus, an example of a task automated by experts. Since automaticity serves as a means of both freeing up cognitive resources for other work functions and improving the speed and accuracy of a task, work systems should facilitate automaticity wherever possible. Because additional stressors in the EMS worker's environment likely contribute to difficulties with drug administration, allowing for basic tasks such as IV injection to be completed as easily as possible would increase the amount of attention and deliberate action devoted to more difficult aspects of the task such as selection of the correct medications or the calculation of dosage information. For example, recent studies on the administration of medication in pediatric cardiac arrests indicate that the range of clinically equivalent medication options for treating hyperkalemia (all acceptable choices within current clinical guidelines) has widely different preparation and administration times due, in great part, to differences in packaging of drugs and manufacturing of products (Arnholt et al., 2015). In situations such as these, an NDM-based approach can be valuable for understanding the factors that drive (or inhibit) effective decision-making as well as the generation of insight into system-level interventions for improving decision-making.

IMPLICATIONS FOR TRAINING

While the design of systems can enable or undermine human decision-making, the expertise of the people making decisions is ultimately at the core, and experience and learning (both formal and informal) determine that expertise (Weaver et al., 2012). These two components—human expertise and system design—are intrinsically linked. For example, clinical guidelines detail how to respond in certain EMS-related situations (e.g., cardiac arrest, trauma), but these guidelines are often ambiguous (Gurses et al., 2008). Working through policies and procedures to find a concrete set of procedures for a given task situation can simplify the training demands placed on people as well as guide more effective system design (i.e., the system needs to support a narrower range of behaviors; Arnholt et al., 2015). In this section, we focus on the developmental mechanisms through which experts are created. Specifically, we provide a brief introduction to four general principles of developing expertise and discuss how these can inform training and development strategies throughout the career of EMS workers.

DELIBERATE AND GUIDED PRACTICE

Experts generally have more experience in a work domain than those at lower levels of proficiency. For example, Breckwoldt et al. (2012) found that experts in prehospital endotracheal intubation (defined by their success rate) performed the task more often than those with less experience. However, it is not just more experience that has been shown to account for differences in expertise but more of a specific kind of experience—deliberate practice. There are four conditions of deliberate practice: repetition of the same or similar task(s), immediate developmental feedback targeted at improvement, a progression that builds on the learners' existing knowledge but pushes them beyond their current performance levels, and motivation on the part of the learner to persist and improve (Ericsson et al., 1993; Krampe & Ericsson, 1996; Ericsson, 2004). Deliberate practice can be challenging to incorporate into practice for domains such as EMS characterized by high production pressure, but it is possible (e.g., through the use of simulations for practicing procedural skills; Lubin & Carter 2009).

SELF-REGULATION

Experts monitor their own performance, diagnose what is effective and ineffective, and adapt their performance processes to achieve higher levels of outcomes. They are better at recognizing their own errors (Glaser, Chi & Farr, 1988) and correcting them. This self-regulation can focus on the external environment, internal cognition processes and affective states, and behavioral performance (Bandura, 1986). In practice, experts frequently achieve this regulation during preplanning activities in preparation for a task, performance control during a task, and postperformance reflective activities (e.g., debriefings; Zimmerman, 2006).

FEEDBACK-SEEKING BEHAVIOR

Feedback is a critical component of the deliberate practice approach, but even outside of these structure learning experiences, experts eagerly seek out feedback on their performance "on the job" (Sonnentag & Kleine, 2000). In domains such as EMS, it is uncommon to accumulate large amounts of time in dedicated training after leaving basic education; therefore, accumulating timely, accurate, and diagnostic (i.e., pinpointing developmental needs) feedback from work activities is critical (Shanteau & Stewart, 1992).

MOTIVATION AND GOAL SETTING

All of the developmental mechanisms described earlier require large amounts of effort, energy, and focus over extended periods (years and even decades). Consequently, motivation is central to most theoretical accounts of expertise development (Ericsson et al., 1993; Sternberg, 1998), and the "rage to master" has been identified as a hallmark of experts (Winner & Drake, 1996). Specifically, four features tend to characterize expert motivation: (1) self-efficacy beliefs, (2) goal orientation, (3) an intrinsic motivation for the task domain, and (4) motivation rooted in a drive for success and excellence, not motivation stemming from a fear of failure (Zimmerman, 2006).

CONCLUDING REMARKS

NDM has proven to be a valuable framework to view performance in complex work domains and generate actionable insights into systems changes—be those the physical environment, devices, the information systems, or training opportunities provided to people. Prehospital EMS is an area that is ripe for the application of NDM. For researchers, it affords a complex, fascinating, and important context of investigation, and for applied professionals, it provides a valuable opportunity to make contributions to improving the safety and quality of care.

REFERENCES

Agarwal, S., Swanson, S., Murphy, A., Yaeger, K., Sharek, P. & Halamek, L.P. (2005). Comparing the utility of a standard pediatric resuscitation cart with a pediatric resuscitation cart based on the Broselow tape: A randomized, controlled, crossover trial involving simulated resuscitation scenarios. *Pediatrics, 116*, e326–e333.

Arnholt, A.M., Duval-Arnould, J.M., McNamara, L.M., Rosen, M.A., Singh, K. & Hunt, E.A. (2015). Comparatively evaluating medication preparation sequences for treatment of hyperkalemia in pediatric cardiac arrest: A prospective, randomized, simulation-based study. *Pediatric Critical Care Medicine, 16*, e224–e230.

Bandura, A. (1986). *Social foundations of thought and action: A social cognitive theory.* Upper Saddle River, NJ: Prentice-Hall.

Breckwoldt, J., Klemstein, S., Brunne, B., Schnitzer, L., Arntz, H.R. & Mochmann, H.C. (2012). Expertise in prehospital endotracheal intubation by emergency medicine physicians: Comparing "proficient performers" and "experts." *Resuscitation, 83*, 434–439.

Cannon-Bowers, J.A., Salas, E. & Pruitt, J.S. (1996). Establishing the boundaries of a paradigm for decision-making research. *Human Factors, 38*, 193–205.

Chase, W.G. & Simon, H.A. (1973). Perception in chess. *Cognitive Psychology, 4*, 55–81.

Cooke, N.J. (1999). Knowledge elicitation. In F. Durso (Ed.), *Handbook of applied cognition* (pp. 479–509). Chichester, UK: Wiley.

Craik, K.J.W. (1943). *The nature of explanation.* Cambridge, MA: Cambridge University Press.

Drews, F.A. & Kramer, H.S. (2012). Human-computer interaction design in health care. In P. Carayon (Ed.), *Handbook of human factors in health care and patient safety* (pp. 265–280). Boca Raton, FL: CRC Press.

Ericsson, K.A. (2004). Deliberate practice and the acquisition and maintenance of expert performance in medicine and related domains. *Academic Medicine, 79*, S70–S81.

Ericsson, K.A. & Charness, N. (1994). Expert performance: Its structure and acquisition. *American Psychologist, 49*, 725–747.

Ericsson, K.A., Krampe, R.T. & Tesch-Römer, C. (1993). The role of deliberate practice in the acquisition of expert performance. *Psychological Review, 100*, 363.

Feufel, M.A., Lippa K.D. & Klein H.A. (2009). Calling 911: Emergency medical services in need of human factors. *Ergonomics in Design, 17*, 15–19.

Flanagan, J.C. (1954). The critical incident technique. *Psychological Bulletin, 51*, 327–358.

Glaser, R., Chi, M.T. & Farr, M.J. (Eds.). (1988). *The nature of expertise.* New York: Lawrence Erlbaum Associates.

Gobet, F. & Simon, H.A. (1996). Templates in chess memory: A mechanism for recalling several boards. *Cognitive Psychology, 31*, 1–40.

Grossman, R., Spencer, J.M. & Salas, E. (2014). Enhancing naturalistic decision making and accelerating expertise in the workplace: Training strategies that work. In S. Highhouse, R.S. Dalal & E. Salas (Eds.), *Judgment and decision making at work* (pp. 277–325). New York: Routledge.

Gunnarsson, B.M. & Stomberg, M.W. (2009). Factors influencing decision making among ambulance nurses in emergency care situations. *International Emergency Nursing, 17*, 83–89.

Gurses, A.P., Seidl, K.L., Vaidya, V., Bochicchio, G., Harris, A.D., Hebden, J. & Xiao, Y. (2008). Systems ambiguity and guideline compliance: A qualitative study of how intensive care units follow evidence-based guidelines to reduce healthcare-associated infections. *Quality and Safety in Health Care, 17*, 351–359.

Hubble M.W., Paschal K.R. & Sanders T.A. (2000). Medication calculation skills of practicing paramedics. *Prehospital Emergency Care, 4*, 253–260.

Klein, G. (1993a). A recognition-primed decision (RPD) model of rapid decision making. In G.A. Klein, J. Orasanu, R. Calderwood & C.E. Zsambok (Eds.), *Decision making in action: Models and methods* (pp. 138–147). Norwood, CT: Ablex.

Klein, G. (1993b). *Naturalistic decision making: Implications for design*. Technical report, Crew Systems Ergonomics Information Analysis Center.

Klein, G. (2008). Naturalistic decision making. *Human Factors, 50*, 456–460.

Klein, G., Phillips, J.K., Rall, E.L. & Peluso, D.A. (2007). A data-frame theory of sense-making. In R.R. Hoffman (Ed.), *Expertise out of context* (pp. 113–155). Mahwah, NJ: Lawrence Erlbaum.

Klein, G.A., Calderwood, R. & MacGregor, D. (1989). Critical decision method for eliciting knowledge. *IEE Transactions on Systems, Man, and Cybernetics, 19*, 462–472.

Klein, G.A. & Crandall, B.W. (1995). The role of mental simulation in naturalistic decision making. In J. Flach, P. Hancock, J. Caird & K. Vicente (Eds.), *The ecology of human–machine systems* (pp. 324–358). Hillsdale, NJ: Lawrence Erlbaum Associates.

Krampe, R.T. & Ericsson, K.A. (1996). Maintaining excellence: Deliberate practice and elite performance in young and older pianists. *Journal of Experimental Psychology: General, 125*, 331.

Lesgold, A.M., Rubinson, H., Feltovich, P.J., Glaser, R., Klopfer, D. & Wang, Y. (1988). Expertise in a complex skill: Diagnosing X-ray pictures. In M.T.H. Chi, R. Gaser & M. Farr (Eds.), *The nature of expertise*. (pp. 311–342). Hillsdale, NJ: Erlbaum.

Lipshitz, R. Klein, G., Orasanu, J. & Salas, E. (2001). Taking stock of naturalistic decision making. *Journal of Behavioral Decision Making, 14*, 331–352.

Lubin, J. & Carter, R. (2009). The feasibility of daily mannequin practice to improve intubation success. *Air Medical Journal, 28*, 195–197.

Miller, G. (1956). The magical number seven, plus or minus two: Some limits on our capacity for processing information. *Psychological Review, 63*, 81–97.

Moors, A. & De Houwer, J. (2006). Automaticity: A theoretical and conceptual analysis. *Psychological Bulletin, 132*, 297–326.

Orasanu, J. (2005). Crew collaboration in space: A naturalistic decision-making perspective. *Aviation, Space, and Environmental Medicine, 76*, B154–B163.

Orasanu, J. & Connolly, T. (1993). The reinvention of decision making. In G.A. Klein, J. Orasanu, R. Calderwood & C.E. Zsambok (Eds.), *Decision Making in Action: Models and Methods* (pp. 3–20). Norwood, NJ: Ablex.

Randel, J.M., Pugh, H.L. & Reed, S.K. (1996). Differences in expert and novice situation awareness in naturalistic decision making. *International Journal of Human-Computer Studies, 45*, 579–597.

Rosen, M.A., Salas, E., Lyons, R. & Fiore, S. (2008). Expertise and naturalistic decision making in organizations: Mechanisms of effective decision making. In G.P. Hodgkinson & W.H. Starbuck (Eds.) *The Oxford handbook of organizational decision making: Psychological and management perspectives* (pp. 211–230). Oxford: Oxford University Press.

Salas, E., Rosen, M.A. & DiazGranados, D. (2010). Expertise-based intuition and decision making in organizations. *Journal of Management, 36*, 941–973.

Schwartz, B., Burgess, R., Craig, A. & Wichman, K. (2006, January). *Human factors field study of equipment and process design and medication safety*. Paper presented at the annual meeting of the National Association of EMS Physicians, Naples, FL.

Shanteau, J. & Stewart, T.R. (1992). Why study expert decision making? Some historical perspectives and comments. *Organizational Behavior and Human Decision Processes, 53*, 95–106.

Sonnentag, S. & Kleine, B.M. (2000). Deliberate practice at work: A study with insurance agents. *Journal of Occupational and Organizational Psychology, 73*, 87–102.

Sternberg, R.J. (1998). Abilities are forms of developing expertise. *Educational Researcher, 27*, 11–20.

Weaver, S.J., Newman-Toker, D.E. & Rosen, M.A. (2012). Reducing cognitive skill decay and diagnostic error: Theory-based practices for continuing education in health care. *Journal of Continuing Education in the Health Professions, 32*, 269–278.

Webber, S.S., Chen, G., Payne, S.C., Marsh, S.M. & Zaccaro, S.J. (2000). Enhancing team mental model measurement with performance appraisal practices. *Organizational Research Methods, 3*, 307–322.

Weick, K.E. (1993). Sensemaking in organizations: Small structures with large consequences. In J.K. Murnighan (Ed.), *Social psychology in organizations: Advances in theory and research* (pp. 10–37). Upper Saddle River, NJ: Prentice Hall.

Winner, E. & Drake, J.E. (1996). The rage to master: The decisive role of talent in the visual arts. In S.B. Kaufman, (Ed.), *The complexity of greatness: Beyond talent or practice* (pp. 333–366). New York: Oxford University Press.

Zimmerman, B.J. (2006). Development and adaptation of expertise: The role of self-regulatory processes and beliefs. In K.A. Ericsson, N. Charness, P.J. Feltovich & R.R. Hoffman (Eds.), *The Cambridge handbook of expertise and expert performance* (pp. 705–722). New York: Cambridge University Press.

Zsambok, C.E. (1997). Naturalistic decision making: Where are we now? In C.E. Zsambok & G. Klein (Eds.), *Naturalistic decision making* (pp. 3–16). Mahwah, NJ: Erlbaum.

5 Stress and Performance in Emergency Medical Services

James L. Szalma

CONTENTS

Stress is ubiquitous in modern work, and its effects can be costly in terms of performance and health as well as in public safety and well-being. There is clear evidence that the stress associated with EMS work incurs such costs (e.g., Cydulka et al., 1989; Marmar et al., 1996; Holland, 2011; Donnelly, 2012; Halpern et al., 2012) and that the stress can be acute or chronic (Halpern & Maunder, 2011; Adriaenssens et al., 2015). In addition, stressful events tend to be characteristics of the specific contexts or situations in which they occur (e.g., the patient, bystanders, dangerous conditions), organizational factors (e.g., staff shortages, lack of training or resources, perceived unfairness in the distribution of responsibilities; Donnelly et al., 2014), and characteristics of the person (e.g., general life or work stress; feelings of helplessness; Halpern et al., 2012). In this chapter, I provide an overview of the stress construct, theories of stress, potential avenues for mitigation, and gaps in our understanding of stress and human performance. The issues discussed are relevant across most work domains but specifically germane to the contextual and task factors that affect the performance and well-being of EMS personnel.

THE STRESS CONSTRUCT: DEFINITION AND THEORY

Current definitions of stress are characterized by cognitive and physiological response to environmental demand (Hockey, 1984, 1986, 1997; Hancock & Warm, 1989). With respect to the former, stress may be conceptualized as "cognitive patterning" that comprises the response to stress that varies across different sources of environmental demands (Hockey & Hamilton, 1983) or as a cognitive appraisal regarding the meaningfulness of the stressor and one's capacity to effectively respond to it (Lazarus & Folkman, 1984). Stress is thus a *relational construct* in that it arises from a transaction between the person and the demands of the environment (Lazarus, 1999; Matthews, 2001). From a transactional perspective, stress is defined as an appraisal by an individual that an event or stimulus is a threat to his/her physical or psychological well-being and that the demands posed by the environment exceed the individual's available resources to effectively cope and respond (Lazarus & Folkman, 1984; Lazarus, 1999; Matthews, 2001). Transactional theory can account for why stress responses vary widely across individuals and contexts. A given stimulus may induce a different stress response in two individuals because of differences in how they appraise the event (i.e., the *meaning* of the event for the well-being of one person may differ from that of another person). However, a particular individual may respond quite differently across different situations because different stimuli have distinct patterns of effect on cognition (Hockey & Hamilton, 1983).

META-THEORETICAL ISSUES

THE RELATIONAL APPROACH: IMPLICATIONS FOR STRESS AS OUTCOME VERSUS CAUSE

In conceptualizing and studying stress, it is important to specify whether one is investigating stress as a causal agent (e.g., how stress impairs performance or health) versus a consideration of stress as an outcome (e.g., the effect of task load on stress response). In essence, stress may be viewed as either a cause or an effect, depending on which portion of the transactional cycle one chooses to examine. From this perspective, cause versus effect is a misleading question because it inappropriately focuses attention on stress as a static psychological construct distinct from the context and frozen in time. A better question is how environmental demands affect an individual's response and how this transaction manifests in terms of cognitive state (Hockey, 1984, 1997, 2003) and adaptation to the environmental demands (Hancock & Warm, 1989).

RELATION OF THE STRESS CONSTRUCT TO FATIGUE AND WORKLOAD

Research on stress, fatigue, and workload comprises distinct but related literatures. There have been attempts at integration in which these phenomena are framed as interacting *biobehavioral states* (Gaillard, 2001) or as *operator functional states*

(Hockey, 2003), and stress, fatigue, and workload have been linked to common theoretical frameworks (e.g., Hancock et al., 2012; Hockey, 2012). However, they are distinct constructs. It is clear that workload can be stressful (Hancock & Warm, 1989; Hancock & Caird, 1993) and that compensating for fatigue can increase stress (Hancock & Verwey, 1997; Desmond & Hancock, 2001). However, the causal interrelationship among these constructs remains to be clearly defined. In many domains, including EMS, all three phenomena affect performance and well-being, and for these applications, the concept of operator functional state proposed by Hockey (2003) may be particularly useful. This concept integrates stress, fatigue, and workload in a common state construct defined in terms of an operator's adaptive capacity (see Hancock & Warm, 1989).

Cognitive–Energetic Perspective

The approach adopted here may be considered an "energetic approach" to human performance (Kahneman, 1973; Hancock & Warm, 1989; Hockey, 1997). In contrast to the unitary arousal theory (Hebb, 1955), the cognitive–energetic perspective views performance not as an outcome of a single arousal mechanism but rather as an outcome of cognitive capacities or resources allocated to tasks that individuals must perform (Kahneman, 1973; Norman & Bobrow, 1975; Wickens, 2002). These resources are energetic in that arousal influences the amount available at a given moment, but they are also structural in that they represent attentional capacity (Hancock & Warm, 1989; Hockey, 1997; for a discussion of energetic and structural metaphors for resources, see Hancock & Szalma, 2007; Szalma et al., 2012). The cognitive–energetic approach views stress as a multidimensional construct composed of affective, cognitive, and energetic components (Matthews et al., 2002).

THEORIES OF STRESS

There are three theories of stress that have been most influential in human factors research on performance in operational settings. These are the dynamic adaptability theory (Hancock & Warm, 1989), the compensatory control model (Hockey, 1997), and the cognitive appraisal theory (Lazarus & Folkman, 1984). The three perspectives are complementary rather than contradictory (Hancock & Szalma, 2007, 2008; Szalma et al., 2012). Theoretical integration is outside the scope of this chapter, but all three theories are compatible with the view that stress occurs when the resources demanded exceed those available for allocation to the task. These mechanisms operate at multiple levels within the organism, ranging from neurological to molar behavior (Matthews, 2001).

Appraisal (Transactional) Theory

One of the most well-known cognitive theories of stress is the transactional theory advocated by Lazarus (Lazarus & Folkman, 1984; Lazarus, 1999). It was conceived as a broad theory applicable to acute and chronic life stress, but Matthews (2001) proposed a transactional perspective that emphasizes the importance of cognitive mechanisms of appraisal and choice of coping strategy for understanding the performance effects of stress. The cognitive, affective, and energetic components of stress, as well as performance outcomes, result from cognitive appraisals and self-regulatory coping mechanisms (see Figure 5.1). These processes are determined jointly by environmental factors and the personality of the individual (Lazarus & Folkman, 1984; Matthews, 2008).

The outcomes of the evaluative processes are patterns of appraisal called "core relational themes" (Lazarus, 1991). For instance, the core relational theme for anxiety is existential threat. Thus, stress occurs when events are appraised as hindering or blocking attainment of their desired outcomes (goals). Three core relational themes have been found to be particularly relevant to responding to stress in a performance context, and they correspond to three dimensions of cognitive state that vary as a function of stress (Matthews et al., 1999, 2002, 2013). These are *task engagement*, which reflects the core relational theme of commitment to effort (toward task goals); *distress*, which reflects the theme of overload of processing capacity; and *worry*, which is related to the theme of self-evaluation (for a more detailed treatment, see Matthews et al., 2002). Several studies have established that manipulation of task demands is associated with predicted changes in task engagement, distress, and worry and that personality traits can predict cognitive states (Matthews et al., 1999, 2013; Szalma, 2009; Szalma & Taylor, 2011).

DYNAMIC ADAPTABILITY THEORY

The dynamic adaptability theory (Hancock & Warm, 1989) asserts that contrary to the unitary arousal theory, in which there is a single point of optimum adaptation to environmental demand, most individuals are able to maintain stability in adaptation across a wide range of stressors. Further, this adaptation occurs as multiple levels (cf. Matthews, 2001), and stress can occur as a result of excessive demands (hyperstress) or an excessive absence of stimulation (hypostress; see Figure 5.2a). Increases in stress in either direction push the organism beyond the threshold of dynamic stability, at which point the organism experiences adaptive failure. However, there are multiple forms of adaptation, including affective ("comfort" in Figure 5.2a), behavioral ("psychological" in Figure 5.2a), and physiological. Failures of the different forms of adaptation occur progressively, such that comfort fails first (e.g., one can experience discomfort under stress but maintain high levels of performance), followed by performance and, finally, failure in physiological adaptation (e.g., loss of consciousness; Harris et al., 2005; Tripp et al., 2009).

Hancock and Warm (1989) also identified the task that a person performs as the most proximal source of stress, and they identified two component dimensions of information in the environment that comprise the demands placed on the individual: information rate and information structure (see Figure 5.2b). The former corresponds to the temporal aspects of a stressor (e.g., rate of events, time pressure), and the latter corresponds to how the information is organized within a given task and reflects primarily a spatial dimension of task demands. Both dimensions represent the information in the environment that individuals evaluate and respond to in order to adapt (or fail to adapt) to the circumstances that confront them.

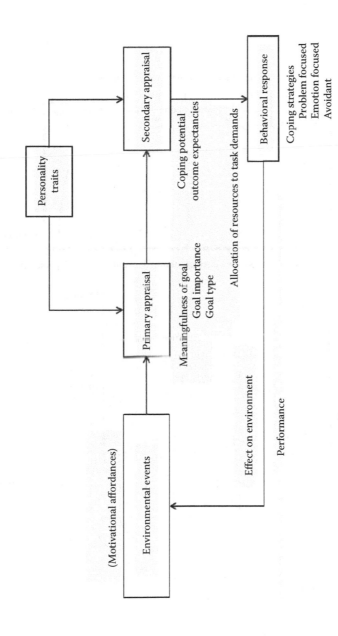

FIGURE 5.1   The transactional model of stress.

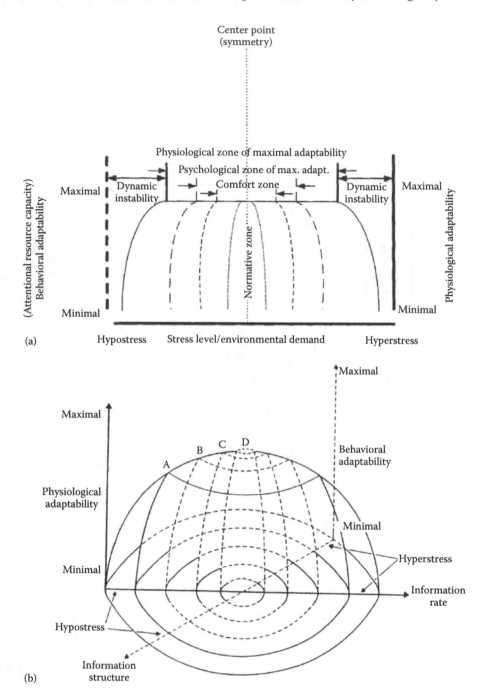

FIGURE 5.2 The dynamic adaptability model of stress (a) and a version of the model in which the stress continuum has been decomposed into its component information dimensions (b). (From Hancock, P.A., and Warm, J.A., *Human Factors*, 31, 519–537, 1989. With permission.)

COMPENSATORY CONTROL MODEL

The dynamic adaptability model accounts for changes in adaptation level as a function of stress imposed, and appraisal theories provide explanations for the role of cognitive mechanisms in stress response, but neither of these theories describes allocation of effort or the mechanisms by which individuals self-regulate behavior (although the latter is implied in both models; Lazarus & Folkman, 1984; Hancock & Warm, 1989). The compensatory control model (CCM; Hockey, 1997, 2003, 2012) was developed to account for the cognitive patterning (Hockey & Hamilton, 1983) of stress and the dependency of operator response on engagement in the task.

The CCM is based on three assumptions: behavior is goal directed, self-regulatory processes control goal states, and regulatory activity has energetic costs (i.e., consumes resources). According to this theory, performance is maintained via a negative feedback self-regulatory control mechanism in which environmental demands are evaluated via an appraisal mechanism, and when a discrepancy is detected, an effort monitor engages a higher-level regulatory loop that provides additional mental resources to accommodate the demand (see Figure 5.3). The lower loop shown in Figure 5.3 (loop A) controls behavior in a relatively automatic fashion and is engaged in contexts in which little effort or mental capacity is required (cf. Schneider & Shiffrin, 1977). However, when the demands are increased beyond the capacity of the lower loop and this increase is detected by the effort monitor, the individual may

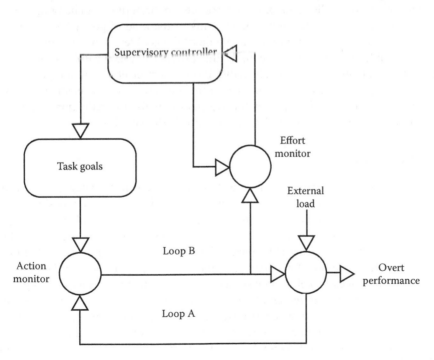

FIGURE 5.3 The CCM. (From Hockey, G.R.J., *Operator Functional State: The Assessment and Prediction of Human Performance Degradation in Complex Tasks,* IOS Press, Amsterdam, 2003. With permission.)

engage the upper loop (loop B) to allocate the necessary resources to maintain task performance. Increases beyond the capacity due to extreme demand or extended periods of work of this upper loop will eventually result in performance failures because the repetitive use of the regulatory process fatigues or strains the compensatory mechanism.

Hockey (1997, 2003) emphasized that the engagement of the upper loop requires expenditure of energy (compensatory effort) and therefore imposes a "hidden cost" to performance, which he referred to as a latent performance decrement. Whether such costs are incurred depends on the decision of an executive function (the supervisory controller in Figure 5.3), which, as seen in Figure 5.3, chooses between two alternatives of increasing effort or adjusting task goals. As Hockey (2003) himself noted, the importance of choice in this model differentiates it from traditional resource theory (Kahneman, 1973) because whether resources are allocated to the task depends on the strategy that the person chooses and is therefore not an automatic compensatory response. If the individual decides to allocate more effort to a task, then increases in demand or prolonged engagement of the compensatory mechanism places him/her in "strain mode," which increases the risk of performance failure. These hidden costs occur in terms of both cognitive states (e.g., distress, worry) and physiological arousal, which occurs as a result of regulatory control activity rather than as a result of changes in the task or other aspects of the environment (Hockey, 1986).

The supervisory system has two strategic responses from which to choose: increasing effort by allocation of more resources to the activity or changing the task goals. The latter modification can be in terms of the kind of goal (abandoning one activity in favor of another) or the magnitude of the goal (e.g., lowering the criterion for acceptable performance). Essentially, this process adjusts the difference between goal state and current state by allocating more effort or changing the goal, represented in Figure 5.3 as the two outputs from the supervisory executive. Effortful coping results in a strain mode of self-regulation, while the reduction of effort or performance goals is a "passive" mode of control (Hockey, 1997).

STRESS MITIGATION

Techniques for stress mitigation tend to fall into three general categories: (i) training on the task, on coping strategies, or on both; (ii) design (or redesign) of the task structure, the interface, or both; and (iii) selecting individuals who are resilient to stress. With respect to the latter approach, I have previously argued (Szalma, 2008, 2009) that measuring individual differences is useful for selection but that a potentially more powerful application of individual differences is to the design of tasks and interfaces as well as in training methods, both adaptive as well as general procedures. Hence, it is recommended here that training and design are the better approaches to mitigating negative stress effects in operational environments, as selection may not always be feasible and consideration of individual differences can improve the design of both technology and of training regimens for all workers in a given domain.

Training for stress may include training individuals to be more resilient by developing more effective task-coping strategies; training on the task so that fewer cognitive capacities are required and, therefore, performance is less vulnerable to the detrimental effects of stress (training to automaticity; Wickens et al., 2013); or training on the task under stressful conditions to "inoculate" the individual to performance-based stress effects (e.g., Varker & Devilly, 2012; for more general treatments outside the EMS domain, see Driskell et al., 2008; Johnston & Cannon-Bowers, 1996).

In EMS operational settings, there are multiple forms and sources of stress; the best approach is likely to be one that includes all three forms of training. Any training procedures for stress should include mastery of effective coping strategies, as maladaptive coping strategies (e.g., avoidant coping; Lazarus & Folkman, 1984; Matthews & Campbell, 1998) are associated with greater stress symptoms in EMS personnel (Holland, 2011). However, further research is needed to determine the best format for training. It is unlikely that traditional workshop-style training will be as effective as training procedures in which the individual has the opportunity to practice the task and coping strategies. However, extensive programmatic research to determine the relative efficacy of different forms of stress training (e.g., in randomized controlled trials) has yet to be done (although for an exception, see Tuckey & Scott, 2014).

Task structure and the interface can each be modified to aid performance in stressful work conditions by following one general but crucial guideline: simplify. This is a guiding principle for human factors in general—to create tasks and interfaces that are as simple as possible to use without sacrificing the necessary functionality. However, simplicity in cognitive demands is particularly important in stressful situations. Tasks that require resource-intensive information processing are the most vulnerable to stress effects (which is why training to automaticity when possible is a useful stress mitigation strategy). Hence, to reduce stress effects, designers should ensure that the tasks require minimal cognitive resources.

There are forms of task structure that may lend themselves to effective support of performance under stress. For instance, Hancock and Szalma (2003) described an approach to preserving performance in stressful environments by using relatively simple perceptual displays (e.g., object or configural displays; see also Szalma, 2011). In the EMS context, this means that displays should be designed so that the information that they present to the operator can be extracted quickly because the elements that convey the information are easily extracted via relatively low-level perceptual mechanisms (e.g., variations in shape, size, or color). In adaptive systems, it may be possible to modify displays for supporting different phases of operator functional state, although use of adaptive displays merits caution as it can potentially introduce system complexity that can undermine functionality.

CONSIDERING THE WHOLE PERSON: STRESS MITIGATION VIA MOTIVATION DESIGN

Task performance does not occur in a vacuum, and human operators are not "pure" information-processing devices without emotions or motives related to their activities. One approach to increasing stress resilience is to structure the tasks, the contexts, and the procedures to be *meaningful* to the person. How one responds to stress and even trauma depends, in part, on the meaning that one ascribes to the experience (Frankl, 1984). If the activity is evaluated as important relative to a person's self-related goals (i.e., goals that are highly valued and integrated with one's sense of self; Ryan & Deci, 2008), then the motivation to perform the activity is strengthened (Sheldon & Kasser, 1995; Deci & Ryan, 2000; Ryan & Deci, 2000, 2001). However, the meaningfulness of interaction with specific tasks involving use of technology, beyond immediate and proximal task goals, has been largely neglected in human factors research (for exceptions, see Hancock, 2009; Szalma, 2014). Szalma (2014) recently proposed a model and set of general guidelines for incorporating motivation theory into human factors design.

SELF-DETERMINATION THEORY

Self-determination theory (SD theory) distinguishes intrinsic motivation from multiple forms of extrinsic motivation (Deci & Ryan, 1985, 2000; Ryan & Deci, 2000, 2008). Intrinsic motivation occurs when the individual experiences inherent satisfaction from the behavior (Ryan & Deci, 2008). A person is "intrinsically motivated only for activities that hold intrinsic interest for them, activities that have the appeal of novelty, challenge, or aesthetic value" (Ryan & Deci, 2000, p. 71). Extrinsic motivation results from an external source, even in cases in which the external value has been internalized. There is strong evidence supporting a distinction between intrinsic and different forms of extrinsic motivation, and there are multiple factors that influence whether behavior is intrinsically or extrinsically motivated (e.g., see Deci et al., 1999).

SD theory assumes that motivation is energized by three innate psychological needs. These are the needs for autonomy (self-determination, an internal perceived locus of causality), competence, and relatedness. These needs and the proposed cognitive evaluation mechanisms of self-regulation are not unique to SD theory. However, the theory is unique in the integration of these constructs from diverse perspectives on motivation as well as in the distinction among qualitatively different forms of motivation. This latter characteristic differs from theories that propose a single self-regulatory (Carver & Scheier, 1998) or goal setting mechanism (Locke & Latham, 1990).

In brief summary, intrinsic motivation is supported by environments that afford autonomous behavior (i.e., choice; Dember & Earl, 1957) and promote development of competence (White, 1959)—and its subjective resultant, self-efficacy (Bandura, 1997)—and relatedness or a sense of meaningful connectedness with other individuals or agents in the environment. Conditions that interfere with or prevent need satisfaction undermine intrinsic motivation.

FORMS OF EXTRINSIC MOTIVATION: THE IMPORTANCE OF AUTONOMY

There are four forms of extrinsic motivation that differ in the level of autonomy experienced by the individual. These are illustrated in Figure 5.4. The lowest level of autonomy is "external regulation," in which behavior is directly regulated by externally determined response contingencies. The next level of extrinsic motivation is

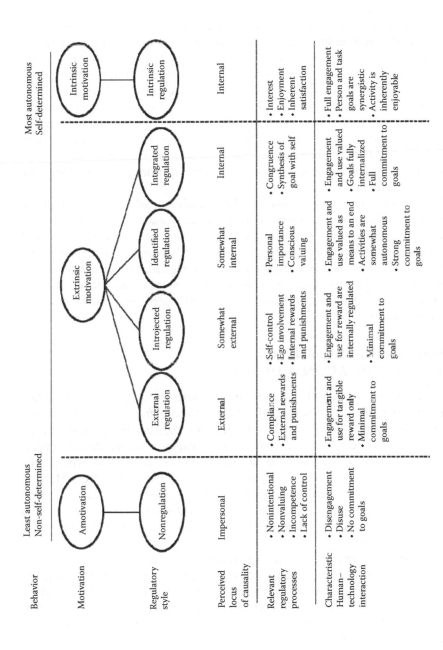

FIGURE 5.4 The forms of motivation and self-regulation in human–technology interaction. (From Szalma, J.L., *Human Factors*, 56, 1453–1471, 2014. With permission.)

introjected regulation, composed of external contingencies that the individual him/ herself regulates. The goal is internalized only to the extent that the person self-regulates their response to the external contingencies. Identified regulation is characterized by external factors that have been internalized by the person and are not linked to specific contingencies. The individual understands the value of the activity for its own sake, as a means to an end, and values it, but the activity is a requirement imposed by an external source. Integrated regulation, the most autonomous form of extrinsic motivation, occurs when a goal is considered important because its value has been integrated (internalized) into the person's self-concept. Goals pursued as identified or integrated regulation are considered *self-concordant*, because the goal-directed activity is congruent with the values and goals that comprise part of the person's self-concept (Sheldon & Elliot, 1999). Integrated regulation is fully autonomous but differs from intrinsic motivation because it results from the utility of the activity for supporting attainment of personally important goals rather than the inherent interest in the activity itself.

Although intrinsic motivation may not be a realistic goal for EMS personnel, identified and integrated regulations are realistic and possible without the need to derive entirely new ways of organizing technological interfaces or the principles of design on which they are based. Creating a work environment, at both the immediate task level and the organizational level that facilitate identified or integrated regulation, will provide the mindfulness of work activity that can confer the resilience to stress individuals in this profession need. The crucial ingredient is provision of real autonomy to workers.

IMPROVING STRESS RESILIENCE THROUGH MOTIVATIONAL DESIGN

One way to increase stress resilience is to create contexts that support self-concordant integrated self-regulation. Szalma (2014) has outlined a framework for applying motivation theory to human factors design (see Figure 5.5), and these guidelines can easily be applied to the design for stressful jobs. In fact, there are particular examples in which EMS stress mitigations have already been developed that incorporate the motivational component (perhaps inadvertently) of performance and stress. Halpern et al. (2009) investigated themes associated with the stress occurring after a critical incident, and they reported that emotional support by one's supervisor and provision of a "timeout" period allow the emotional arousal that occurred immediately after the incident to dissipate.

The two interventions recommended by Halpern et al. (2009), emotional support from the supervisor and a timeout period soon after the stressful event, are likely effective because they support the autonomy and relatedness needs of the individual. In fact, the emotional support recommended by Halpern et al. (2009, p. 146) "consists of: acknowledgement of the incident as critical, valuing the work done by the emergency medical technician (EMT), concern about the well-being of the EMT, willingness to listen and to offer material help." All of these techniques are recommended ways to promote a work environment that supports autonomy and relatedness (Gagne et al., 1997; Gagne & Deci, 2005; Stone et al., 2009).

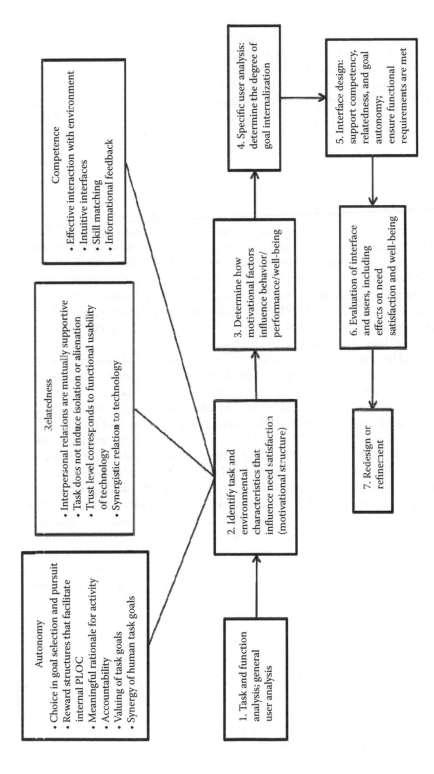

FIGURE 5.5 Guidelines for motivational design in human factors and ergonomics. PLOC, perceived locus of causality. (From Szalma, J.L., *Human Factors*, 56, 1453–1471, 2014. With permission.)

These approaches can reduce occupational exhaustion and enhance commitment (Fernet et al., 2012), possible by facilitating stronger personal resources to respond to job demands (Van den Broeck et al., 2011). Facilitating need satisfaction has been shown to reduce the effects of poor job resources (e.g., limited physical and organizational resources for accomplishing job-related tasks) on exhaustion (Van den Broeck et al., 2008) and to enhance psychological health (Boudrias et al., 2011). Implementing these techniques in EMS organizations should have beneficial effects in terms of performance and stress mitigation, as well as in the engagement and commitment of EMS personnel.

GAPS IN UNDERSTANDING

There are both theoretical and practical gaps in understanding that should be addressed to further advance the field of stress and performance. The central theoretical gap is the refinements of the stress construct itself. Stress is a functional state with biological, cognitive, and emotional components (Gaillard, 2001). Research efforts should therefore be directed toward a further exploration of the intersections of these elements and of how these interactions affect manifestations of stress response. Accomplishing this will require refinement of currently vague constructs such as cognitive resources, as well as more precise models of cognitive appraisal and of how individuals evaluate goal progress.

A second theoretical gap is in our collective understanding of how individual differences in personality and in cognitive abilities (or skills) affect the experience and performance under stress. Personality has been examined extensively in stress and health research (e.g., for a general review, see Matthews et al., 2009), but its application to cognitive performance under stress, particularly in operational settings, has been somewhat limited. This is likely due to long-standing divisions between disciplines of psychology (Cronbach, 1957, 1975), although there have been efforts to include individual differences in stress and performance research (Hockey, 1986; Matthews et al., 2000, 2009; Szalma, 2008). Further, guidelines for incorporating individual differences into human factors design processes have been proposed (Szalma, 2009), and these can be applied to the design of tasks, interfaces, and training procedures described earlier in this chapter. Motivational design has the potential to ameliorate stress by providing experiences of autonomy, competence, and relatedness that provide meaning to a task (Sheldon et al., 2004) and energize behavior (Nix et al., 1999; Moller et al., 2006; Ryan & Deci, 2008). However, empirical research on stress and performance has yet to fully exploit these methods for the design of real-world operational environments.

CONCLUSIONS

Stress is ubiquitous and it is likely to continue to pose constraints on effective performance in operational settings. Given the literal life-and-death situations faced by EMS personnel, there will always be some degree of stress for medical service providers. Hence, mitigation strategies that involve removal of the source of the stress (e.g., by eliminating noise or controlling the ambient temperature) are unlikely to be effective in the EMS context because stress is an inherent component of the operational environment.

The best approach to confronting stress is therefore to develop ways to reduce its effects on performance and well-being. The best approaches are training and the design of the environment, specifically the task, the interface, and the motivational structures. Stress for EMS personnel will never be completely eliminated, but application of stress theory and research and the mitigation strategies derived from them offer effective ways to manage the stress associated with work in emergency medicine.

REFERENCES

Adriaenssens, J., De Gucht, V. & Maes, S. (2015). Causes and consequences of occupational stress in emergency nurses, a longitudinal study. *Journal of Nursing Management, 23*, 346.

Bandura, A. (1997). *Self-efficacy: The exercise of control.* New York: W.H. Freeman and Company.

Boudrias, J.S., Desrumaux, P., Gaudreau, P., Nelson, K., Brunet, L. & Savoie, A. (2011). Modeling the experience of psychological health at work: The role of personal resources, social-organizational resources, and job demands. *International Journal of Stress Management, 18*, 372–395.

Carver, C.S. & Scheier, M.F. (1998). *On the self-regulation of behavior.* New York: Cambridge University Press.

Cronbach, L.J. (1957). The two disciplines of scientific psychology. *American Psychologist, 12*, 671–684.

Cronbach, L.J. (1975). Beyond the two disciplines of scientific psychology. *American Psychologist, 30*, 116–127.

Cydulka, R.K., Lyons, J., Moy, A., Shay, K., Hammer, J. & Mathews, J. (1989). A follow-up report of occupational stress in urban EMT-paramedics. *Annals of Emergency Medicine, 18*, 1151–1156.

Deci, E.L. & Ryan, R.M. (1985). *Intrinsic motivation and self-determination in human behavior.* New York: Plenum Press.

Deci, E.L. & Ryan, R.M. (2000). The "what" and "why" of goal pursuits: Human needs and the self-determination of behavior. *Psychological Inquiry, 11*, 227–268.

Deci, E.L., Ryan, R.M. & Koestner, R. (1999). A meta-analytic review of experiments examining the effects of extrinsic rewards on intrinsic motivation. *Psychological Bulletin, 125*, 627–668.

Dember, W.N. & Earl, R.W. (1957). Analysis of exploratory, manipulatory, and curiosity behaviors. *Psychological Review, 64*, 91–96.

Desmond, P.A. & Hancock, P.A. (2001). Active and passive fatigue states. In P.A. Hancock & D.A. Desmond (Eds.), *Stress, workload, and fatigue* (pp. 455–465). Mahwah, NJ: Erlbaum.

Donnelly, E. (2012). Work-related stress and posttraumatic stress in emergency medical services. *Prehospital Emergency Care, 16*, 76–85.

Donnelly, E.A., Chonody, J. & Campbell, D. (2014). Measuring chronic stress in the emergency medical services. *Journal of Workplace Behavioral Health, 29*, 333–353.

Driskell, J.E., Salas, E., Johnston, J.H. & Wollert, T.N. (2008). Stress exposure training: An event-based approach. In P.A. Hancock & J.L. Szalma (Eds.), *Performance under stress* (pp. 271–286). Aldershot, UK: Ashgate.

Fernet, C., Austin, S. & Vallerand, R.J. (2012). The effects of work motivation on employee exhaustion and commitment: An extension of the JD-R model. *Work and Stress, 26*, 213–229.

Gagne, M. & Deci, E.L. (2005). Self-determination theory and work motivation. *Journal of Organizational Behavior*, 26, 331–362.

Gagne, M., Senecal, C.B. & Koestner, R. (1997). Proximal job characteristics, feelings of empowerment, and intrinsic motivation: A multidimensional model. *Journal of Applied Social Psychology*, 27, 1222–1240.

Gaillard, A.W.K. (2001). Stress, workload, and fatigue as three biobehavioral states: A general overview. In P.A. Hancock & P.A. Desmond (Eds.), *Stress, workload, and fatigue* (pp. 623–639). Mahwah, NJ: Erlbaum.

Halpern, J., Gurevich, M., Schwartz, B. & Brazeau, P. (2009). Interventions for critical incident stress in emergency medical services: A qualitative study. *Stress and Health: Journal of the International Society for the Investigation of Stress*, 25, 139–149.

Halpern, J. & Maunder, R.G. (2011). Acute and chronic workplace stress in emergency medical technicians and paramedics. In J. Langan-Fox & C.L. Cooper (Eds.), *Handbook of stress in the occupations* (pp. 135–156). Northampton, MA: Edward Elgar Publishing.

Halpern, J., Maunder, R.G., Schwartz, B. & Gurevich, M. (2012). The critical incident inventory: Characteristics of incidents which affect emergency medical technicians and paramedics. *Biomed Central Emergency Medicine*, 12, 10.

Hancock, P.A. & Caird, J.K. (1993). Experimental evaluation of a model of mental workload. *Human Factors*, 35, 413–429.

Hancock, P.A., Desmond, P.A. & Matthews, G. (2012). Conceptualizing and defining fatigue. In G. Matthews, P.A. Desmond, C. Neubauer & P.A. Hancock (Eds.), *The handbook of operator fatigue* (pp. 63–73). Burlington, VT: Ashgate.

Hancock, P.A. & Szalma, J.L. (2003). Operator stress and display design. *Ergonomics in Design*, 11, 13–18.

Hancock, P.A. & Szalma, J.L. (2007). Stress and neuroergonomics. In R. Parasuraman & M. Rizzo (Eds.), *Neuroergonomics: The brain at work* (pp. 195–206). Oxford: Oxford University Press.

Hancock, P.A. & Szalma, J.L. (2008). Stress and performance. In P.A. Hancock & J.L. Szalma (Eds.), *Performance under stress* (pp. 1–18). Aldershot, UK: Ashgate.

Hancock, P.A. & Verwey, W.B. (1997). Fatigue, workload, and adaptive driver systems. *Accident Analysis & Prevention*, 4, 495–506.

Hancock, P.A. & Warm, J.S. (1989). A dynamic model of stress and sustained attention. *Human Factors*, 31, 519–537.

Harris, W.C., Hancock, P.A. & Harris, S.C. (2005). Information processing changes following extended stress. *Military Psychology*, 17, 15–128.

Hebb, D.O. (1955). Drives and the C.N.S. (conceptual nervous system). *Psychological Review*, 62, 243–254.

Hockey, R. (1984). Varieties of attentional state: The effects of environment. In: D.R. Davies & R. Parasuraman (Eds.), *Varieties of Attention* (pp. 449–483). New York: Academic Press.

Hockey, G.R.J. (1986). A state control theory of adaptation and individual differences in stress management. In G.R.J. Hockey, A.W.K Gaillard & M.G.H. Coles (Eds.), *Energetic aspects of human information processing* (pp. 285–298). Leiden, Netherlands: Nijhoff.

Hockey, G.R.J. (1997). Compensatory control in the regulation of human performance under stress and high workload: A cognitive–energetical framework. *Biological Psychology*, 45, 73–93.

Hockey, G.R.J. (2003). Operator functional state as a framework for the assessment of performance. In A.W.K. Gaillard, G.R.J. Hockey & O. Burov (Eds.), *Operator functional state: The assessment and prediction of human performance degradation in complex tasks* (pp. 8–23). Amsterdam: IOS Press.

Hockey, G.R. (2012). A motivational control theory of cognitive fatigue. In P.L. Ackerman (Ed.), *Cognitive fatigue: Multidisciplinary perspectives on current research and future applications* (pp. 167–187). Washington, DC: American Psychological Association.

Hockey, R. & Hamilton, P. (1983). The cognitive patterning of stress states. In R. Hockey (Ed.), *Stress and fatigue in human performance* (pp. 331–362). Chichester, UK: Wiley.

Holland, M. (2011). The dangers of detrimental coping in emergency medical services. *Prehospital Emergency Care, 15*, 331–337.

Johnston, J.H. & Cannon-Bowers, J.A. (1996). Training for stress exposure. In J.E. Driskell & E. Salas (Eds.), *Stress and human performance* (pp. 223–256). Mahwah, NJ: Erlbaum.

Kahneman, D. (1973). *Attention and effort.* Englewood Cliffs, NJ: Prentice Hall.

Lazarus, R.S. (1991). *Emotion and adaptation.* Oxford: Oxford University Press.

Lazarus, R.S. (1999). *Stress and emotion: A new synthesis.* New York: Springer.

Lazarus, R.S. & Folkman, S. (1984). *Stress, appraisal, and coping.* New York: Springer.

Locke, E.A. & Latham, G.P. (1990). *A theory of goal setting & task performance.* Englewood Cliffs, NJ: Prentice Hall.

Marmar, C.R., Weiss, D.S., Metzler, T.J., Ronfeldt, H.M. & Foreman, C. (1996). Stress responses of emergency services personnel to the Loma Prieta earthquake interstate 880 freeway collapse and control traumatic incidents. *Journal of Traumatic Stress, 9*, 63–85.

Matthews, G. (2001). Levels of transaction: A cognitive science framework for operator stress. In P.A. Hancock & P.A. Desmond (Eds.), *Stress, workload, and fatigue* (pp. 5–33). Mahwah, NJ: Erlbaum.

Matthews, G. (2008). Personality and information processing: A cognitive-adaptive theory. In G. Matthews, G.J. Boyle & D.H. Saklofske (Eds.), *Handbook of personality theory and assessment: Volume 1: Personality theories and models* (pp. 56–79). Thousand Oaks, CA: Sage.

Matthews, G. & Campbell, S.E. (1998). Task-induced stress and individual differences in coping. *Proceedings of the Human Factors and Ergonomics Society, 42*, 821–825.

Matthews, G., Campbell, S.E., Falconer, S., Joyner, L.A., Huggins, J., Gilliland, K., Grier, R., Warm, J.S. (2002). Fundamental dimensions of subjective state in performance settings: Task engagement, distress, and worry. *Emotion, 2*, 315–340.

Matthews, G., Davies, D.R., Westerman, S.J. & Stammers, R.B. (2000). *Human performance: Cognition, stress, and individual differences.* London: Psychology Press.

Matthews, G., Deary, I.J. & Whiteman, M.C. (2009). *Personality traits* (3rd ed.). Cambridge, UK: Cambridge University Press.

Matthews, G., Joyner, L., Gilliland, K., Campbell, S., Falconer, S. & Huggins, J. (1999). Validation of a comprehensive stress state questionnaire: Towards a state "big three"? In I.J. Deary, I. Mervielde, F. De Fruyt & F. Ostendorf (Eds.), *Personality psychology in Europe*, vol. 7 (pp. 335–350). Tilburg, the Netherlands: Tilburg University Press.

Matthews, G., Szalma, J.L., Panganiban, A., Neubauer, C. & Warm, J.S. (2013). Profiling task stress with the Dundee Stress State questionnaire. In L. Cavalcanti & S. Azevedo (Eds.), *Psychology of stress: New research* (pp. 49–91). Hauppauge, New York: Nova Publishers.

Moller, A.C., Deci, E.L. & Ryan, R.M. (2006). Choice and ego-depletion: The moderating role of autonomy. *Personality and Social Psychology Bulletin, 32*, 1024–1036.

Nix, G.A., Ryan, R.M., Manly, J.B. & Deci, E.L. (1999). Revitalization through self-regulation: The effects of autonomous and controlled motivation on happiness and vitality. *Journal of Experimental Social Psychology, 35*, 266–284.

Norman, D. & Bobrow, D. (1975). On data-limited and resource-limited processing. *Journal of Cognitive Psychology, 7*, 44–60.

Ryan, R.M. & Deci, E.L. (2000). Self-determination theory and the facilitation of intrinsic motivation, social development, and well-being. *American Psychologist, 55*, 66–78.

Ryan, R.M. & Deci, E.L. (2008). From ego depletion to vitality: Theory and findings concerning the facilitation of energy available to the self. *Social and Personality Psychology Compass, 2/2*, 702–717.

Schneider, W. & Shiffrin, R.M. (1977). Controlled and automatic human information processing I: Detection, search, and attention. *Psychological Review, 84*, 1–66.

Sheldon, K.M. & Elliot, A.J. (1999). Goal striving, need satisfaction, and longitudinal well-being: The self-concordance model. *Journal of Personality & Social Psychology, 76*, 482–497.

Sheldon, K.M. & Kasser, T. (1995). Coherence and congruence: Two aspects of personality integration. *Journal of Personality and Social Psychology, 68*, 531–543.

Sheldon, K.M., Ryan, R.M., Deci, E.L. & Kasser, T. (2004). The independent effects of goal contents and motives on well-being: It's both what you pursue and why you pursue it. *Personality and Social Psychology Bulletin, 30*, 475–486.

Stone, D.N., Deci, E.L. & Ryan, R.M. (2009). Beyond talk: Creating autonomous motivation through self-determination theory. *Journal of General Management, 34*, 75–91.

Szalma, J.L. (2008). Individual differences in stress reaction. In P.A. Hancock & J. Szalma (Ed.), *Performance under stress* (pp. 323–357). Aldershot, UK: Ashgate.

Szalma, J.L. (2009). Individual differences: Incorporating human variation into human factors/ergonomics research and practice. *Theoretical Issues in Ergonomics Science, 10*, 381–397.

Szalma, J.L. (2011). Workload and stress in vigilance: The impact of display format and task type. *American Journal of Psychology, 124*, 441–454.

Szalma, J.L. (2014). On the application of motivation theory to human factors/ergonomics: Motivational design principles for human–technology interaction. *Human Factors, 56*, 1453–1471.

Szalma, J.L., Hancock, G.M. & Hancock, P.A. (2012). Task loading and stress in human-computer interaction: Theoretical frameworks and mitigation strategies. In J.A. Jacko (Ed.), *The human–computer interaction handbook: Fundamentals, evolving technologies, and emerging applications* (3rd ed., pp. 55–75). Boca Raton, FL: CRC Press.

Szalma, J.L. & Taylor, G.S. (2011). Individual differences in response to automation: The big five factors of personality. *Journal of Experimental Psychology: Applied, 17*, 71–96.

Szalma, J.L. & Teo, G.W.L. (2012). Spatial and temporal task characteristics as stress: A test of the dynamic adaptability theory of stress, workload, and performance. *Acta Psychologica, 139*, 471–485.

Tripp, L.D., Warm, J.S. & Matthews, G. (2009). On tracking the course of oxygen saturation and pilot performance during gravity-induced loss of consciousness. *Human Factors, 51*, 775–784.

Tuckey, M.R. & Scott, J.E. (2014). Group critical incident stress debriefing with emergency services personnel: A randomized controlled trial. *Anxiety, Stress & Coping, 27*, 38–54.

Van den Broeck, A., Vansteenkiste, M., De Witte, H. & Lens, W. (2008). Explaining the relationships between job characteristics, burnout, and engagement: The role of basic psychological need satisfaction. *Work and Stress, 22*(3), 277–294.

Van den Broeck, A., Van Ruysseveldt, J., Smulders, P. & De Witte, H. (2011). Does an intrinsic work value orientation strengthen the impact of job resources? A perspective from the Job Demands-Resources Model. *European Journal of Work and Organizational Psychology, 20*(5), 581–609.

Varker, T. & Devilly, G.J. (2012). An analogue trial of inoculation/resilience training for emergency services personnel: Proof of concept. *Journal of Anxiety Disorders, 26*, 696–701.

White, R.W. (1959). Motivation reconsidered: The concept of competence. *Psychological Review, 66*, 297–333.

Wickens, C.D. (2002). Multiple resources and performance prediction. *Theoretical Issues in Ergonomics Science, 3*, 159–177.

Wickens, C.D., Hollands, J.G., Banbury, S. & Parasuraman, R. (2013). *Engineering psychology and human performance* (4th ed.). Boston: Pearson.

6 Expertise and Decision-Making in Emergency Medical Services

Stuart Donn

CONTENTS

INTRODUCTION

In the context of EMS practice, the characteristics identified with the naturalistic decision-making (NDM) enterprise are clearly applicable. Time pressures, high stakes, competing priorities, incomplete information, changing conditions, and vague goals can all be applied to the context in which paramedics operate. Their decision-making is bounded by two broad frameworks: first, situational awareness and, second, patient treatment.

The first, situation awareness (SA), is a necessary but not sufficient condition for good decision-making. The second, seen as primary in the EMS role, is to provide appropriate patient treatment within the scope of practice that delimits paramedic care. As part of the healthcare team, paramedics have the same aim as physicians—do no harm. But there are differences in the decision-making that the two practitioners engage in. Both kinds of decision-making are based on the expertise of their specific practices. For example, paramedics do not need to make a definitive diagnosis as a physician does, nor is it even possible most of the time. The paramedic diagnosis is sufficient when the decision can be made as to what they can do in the field setting, within their scope of practice, given considerations to their differential considerations and which of those represents the biggest life threat.

The constraints imposed by the field setting include limitations on resources and imposition of physical elements. The resource limitation may be an aspect such as a definitive diagnostic procedure—the X-ray or ultrasound identification of specific injuries. These will not be available before patient extrication from an automobile crash. What can be established is that there are injuries, and treatment can be provided for some of the injuries, but for others, the definitive treatment is in a hospital setting.

In other instances, even with the appropriate diagnosis of the issue, the scope of practice limits what the paramedic can actually do. At various levels of licensure, IV treatment can be offered or the patient intubated or certain drugs administered. But not all paramedic license levels have the same capability to engage in these interventions. The result is that the EMS decisions are bounded by both the availability of resources and the limitations on their particular practice.

The emerging aspects involved in expert decision-making in EMS reflect a number of changes that are currently taking place in paramedic practice. The historical role for EMS evolved from the ambulance service provided, in many instances, as a sideline to other ventures, such as funeral services. A seminal document establishing the creation of EMS systems in the United States (National Academy of Sciences, 1966) provides a description of the emergent situations to which the response was the development of a specific curriculum to provide training for responders to trauma. Other descriptions of early ambulance services, such as provision of medical treatment during wars and following disasters, provide a picture of relatively untrained individuals offering services to patients in very needy situations.

The lack of equipment and training in the past minimized what kind of treatments could be provided. The expert decision-making required in these situations was limited—relegated to mostly identifying the fastest route to the source of care. With changes in thinking about the kind of care that paramedics could provide also came the need for changes in the expert decision-making that were required. As the remainder of this chapter details, the continuation of that trend has invoked increasing need for competence not only in medical treatment skills but also in decision-making skills. Equipping today's EMS practitioners with those needed attributes is an increasing challenge.

GENERAL ISSUES REGARDING EXPERTISE IN EMS

Paramedic practice is consistent with the premises of NDM. The parameters that influence a decision in the high-stress, high-stakes, and time-pressured circumstances of EMS practice preclude the possibility of applying an analytic method to making the decision. The factors involved are illustrated in Figure 6.1, which identifies specific confounding elements.

The early work in the NDM field recognized that the analysis of decision-making in laboratory settings, with contrived experiments, was not an accurate reflection of how practitioners actually made decisions in the real world. The development of the recognition-primed decision (RPD) model (Klein, 1998, 2008; Ross et al., 2004) was a result of taking the study of decision-making to the area of practice. There has been some work done on experts within EMS and their decision-making, as distinct from novices (Ryan and Halliwell, 2012).

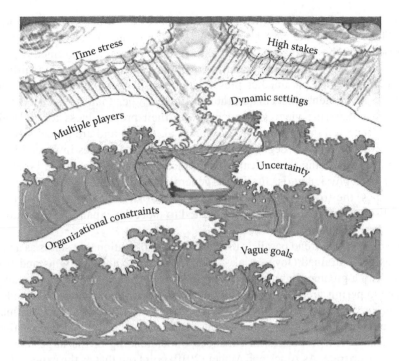

FIGURE 6.1 Confounding elements. (From Klein, G.A. et al., 1993, *Decision making in Action: Models and Methods (Cognition and Literacy).* Praeger.)

Much of the work done on decision-making in healthcare has been associated with physician decision-making, based on their expertise. There has been a transfer of the findings, at least to an extent, to the decision-making of EMS practitioners. The comparisons are reasonable—particularly for the expertise-based decision-making of emergency physicians and EMS paramedics.

ROLE EVOLUTION IN EMS

There are some issues that confound aspects of decision-making for EMS. One of the common aspects facing many professions is the demographic shift, as individuals with accumulated experience are retiring, and the newer generation of practitioners has less experience to draw upon for their decision-making. In the provision of air transport, the patient profile gives insight into the complex presentation of this patient population (Andrew et al., 2015) and the challenges of providing safe treatment for this group can be enhanced by evidence-based guidelines (Thomas et al., 2013). In the absence of "corporate memory" in these complex specialized facets of practice, such reliance on protocols is reasonable. This has been recognized as a significant factor which, when combined with the changing role of technology and complexity of practice, necessitates a closer look at accelerating the acquisition of expertise (Hoffman et al., 2014).

One of the trends in paramedic education is toward a degree-based qualification. This has been the case for some jurisdictions, mainly in English-speaking countries, but the increasing educational requirements for paramedic practice seem more broadly based. The perceived differences between degree-qualified and vocationally trained paramedics present some challenges (O'Meara et al., 2015), but the main contention seems on the skill level rather than on decision-making ability. The differing views of students as to their preparedness for clinical settings are reflected in the study of Hickson et al. (2015) and are potentially a consequence of the move to degree qualifications. The differences in perception may be construed as being theoretical versus practical experience. More broadly, the education of physicians is, first, as a generalist followed by specialization. For paramedics, the trend has been, first, as a specialist, now being followed by the generalist aspect as the education is adjusted to meet the expanding role into community first-line healthcare provision.

The educational trend parallels the increasing expectation that status as a profession has been attained (Joyce et al., 2009; O'Meara, 2009; Williams and Brown, 2010). With a greater alignment to a "professional" designation comes a change in practice to permit greater autonomy in decision-making (e.g., a shift to guidelines from protocols in some areas), as well as broader demands for decision-making when the breadth of practice has increased. There are additional expectations of professionals with respect to an enhanced understanding of the larger context in which they operate. As Atack and Maher (2010) point out that as the expected alignment of EMS with the health sector continues, there will be a need to adjust the EMS education to include increased emphasis on clinical decision-making. However, the differences in the roles, beyond those of scope of practice, include elements such as the diagnosis scope—for physicians, a definitive view, and for paramedics, a diagnosis bounded by what they can do within their scope of practice. Physicians have access to a full range of equipment, tests, and assistance, while paramedics are bounded by the field context—limited in equipment, diagnostic procedures, and availability of assistance.

Thus, the expertise on which paramedics rely is informed by a knowledge translation from medical research to the bounded circumstances of their practice (Bigham and Welsford, 2015). As these authors point out, the translation is not always successful, and a thorough analysis of hospital-based therapeutic success needs to also include the recognition of the different mobile environments of EMS practice. Understanding that the ambulance environment is not simply an extension of the ER means that one cannot assume that what is good in the ER is necessarily good in the field. Similarly, the relationship between the expanded scope of practice for EMS practitioners and the potential impact on patient safety can be mitigated by an increased focus on the decision-making of paramedics, but the research has yet to confirm whether there are negative impacts on patient safety (O'Hara et al., 2015).

Some approaches to instilling the required skills in critical thinking for EMS practitioners can be deduced from the proposal of Facione and Facione (2008). They concluded that the potential for overconfidence errors is acknowledged as a reason to focus on developing suitable critical thinking skills. This element of risk, perhaps as found in the old adage "a little knowledge is a dangerous thing," is echoed in the discussion of medical errors (e.g., Kruger and Dunning, 1998; Berner and Graber, 2008; Croskerry and Norman, 2008; Graber, 2009; Phua and Tan, 2013). Another potential approach is in the development of heuristics that would provide guidance for decision-making in healthcare (Gigerenzer and Gaissmaier, 2011). The appeal of heuristics to improve expert decision-making in EMS is the saving of effort and time, important considerations in their high-stakes, time-pressured work setting. However, the trade-off that frequently accompanies heuristics is that of accuracy over effort reduction. With the emerging increased scope of practice for EMS, the default option of transport for the patient becomes less of an option, and the need for both accuracy and reduced effort in decision-making is necessary.

SPECIFIC ASPECTS OF EXPERTISE FOR EMS

As the importance of decision-making and expertise in EMS continues to expand, so too does the notion of professionalism, and the emergence of increased focus on degree programs in EMS reflects this change. While the concept of increased education for practitioners is sound, the implementation of university-educated graduates into the existing workforce is not entirely smooth (O'Meara 2009, 2011; O'Meara and Grbich, 2009; O'Meara et al., 2015). One element of disagreement is over "book learning" as opposed to "doing the work." While there may be echoes of the 10,000 hours (Ericsson, 1996) in the acquisition of expertise at the base of this disagreement, a question remains as to what skill set is required to meet the expanding scope, increased trend to EMS as a profession, and the increased need for sound decision-making in the practice of EMS.

Woolard's (2012) review of professionalism in UK paramedic practice also raises the notion of "systematic learning from untoward incidents." Here, a basic issue would appear to be identified in the reports of Hobgood et al. (2004) that error identification and reporting are a challenge for all elements of the emergency department (ED) care delivery team. As the role of paramedics is expanded, such disclosures would be a necessary part of individual continuing education identified as part of professional practice. While the capacity to identify and report errors seems available (Hobgood et al., 2006), the application in practice will require additional education and support.

There continues to be a dichotomy in terms of the placement of EMS—are they a public safety service or part of the healthcare system? The question may be framed as "the emergency part of the healthcare system or the health part of the emergency system." For those organizations in which the trend has been to incorporate EMS as part of the healthcare system, reflecting the expansion of the EMS role (O'Hara et al., 2015), there is a recognition that practitioners do not see themselves as full members of the healthcare system, nor are they recognized as full partners by current healthcare practitioners. For the interprofessional education required to enhance functioning in this expanded role, some research indicates that there is readiness on the part of students for interprofessional learning (Williams & Webb, 2015). The authors make reference to other studies on the implementation of interprofessional learning that confirm the need for clear objectives with supporting activities to consolidate the concept. And, as will be noted later, the use of simulation-based activities may be of particular value in this arena.

The expansion of paramedic practice into the area identified as community paramedic appears promising. This appellation has almost as many versions as the communities in which the paramedics serve, and the EPIC study (Expanding Paramedicine in the Community, Drennan et al., 2014) is one attempt to clarify the effectiveness of the community paramedic concept. Others, such as Craig (2015a,b), suggest that there may be alternate approaches to providing care in the community and advocate further study to identify and adopt the better strategy.

SPECIFIC INSTANCE OF WORKING ON EXPERTISE IN SA

In 2007, a tragic accident occurred. Paramedics were called to a mine pumping station aboveground that, unknown to them, was oxygen depleted. The shed had been regularly monitored for a period of 5 years with no indication of a safety issue. The ambulance service was called when the engineer responsible for monitoring the station did not return from his operational check. The first paramedic entered the shed, spotted the downed engineer, and quickly lost consciousness. The second paramedic entered the shed to rescue his partner and was also rendered unconscious. Fortunately, the other paramedics had the good sense not to enter the shed; they called for additional resources. Safety reviews following this accident generated several recommendations to the British Columbia Ambulance Service. Some of these recommendations focused on operational changes to enhance identification of high-risk situations. Some addressed communication issues, specifically on the issue of interagency cooperation. The report from the Ministry of Mines indicated that the circumstances leading to the event were exceedingly rare, and the likelihood of prior identification of the specific risk was small. This incident led to a set of recommendations dealing with the provision of training and education for paramedics to prevent similar incidents in the future.

Discussion within the British Columbia Ambulance Service attempted to match an existing program with the identified needs within the service to no avail. The existing courses tended to have scene hazard recognition as a core element, prescribing specific approaches to known hazardous situations (e.g., drug labs, specific industrial settings, and particular environmental dangers) but providing little in terms of dealing with novel situations (personal communication Toronto EMS, Seattle Fire, and other agencies). The consensus was that a program designed to provide education that would enable paramedics to make informed decisions in high-risk situations that were novel was potentially of greater benefit to their safety than a course which cataloged dangerous contexts and provided specific approaches for them.

Two relevant aspects were established in reviewing the literature pertaining to SA (Endsley, 1995) and decision-making (Klein, 1998). First, the concepts of SA had identified elements that were applicable to paramedic practice although they had not been derived from specific study of paramedics. Aspects such as tunneling, information overload, and lack of recognition of changes to relevant factors were applied for paramedic practice both to the patient being treated and to the context in which the patient was being treated, a potentially hazardous one. Later work (Hamid et al., 2009) is supportive of this approach.

Second, the identification of factors for NDM readily described circumstances in which prehospital practice takes place. The concepts of competing goals, incomplete information, high risk, multiple players, changing conditions, irreducible uncertainty, and organizational constraints all pertained to paramedic practice. The Situation Awareness for Emergency Responders (SAFER) course that was developed utilized the RPD model (Klein, 1998, 2008) and had as a major objective providing a guide for paramedics to use in making safe decisions in their practice.

Initially, participants were given 12 scenarios and asked to indicate which of four possible responses they would choose and which response they believed was what the organization would wish them to choose. The scenarios had been developed by practicing paramedics as were the options for action and reasonably reflected situations that could be encountered in practice. The options were ranked by a group of experienced practitioners from riskiest to safest.

Participants were also given a 30-question survey on their perception of organizational commitment to safety, covering three areas—organizational attitude, aspects of patient safety, and aspects of paramedic safety. The survey was based on the University of Texas Health Sciences organizational safety survey and an instrument used by an Australian airline for similar purposes.

The course consisted of an introduction to SA, the aspects of common ground, the importance of feedback, and how the RPD model illustrates the approaches taken to decision-making in a circumstance embodying the elements of NDM. Several decision-making exercises were carried out during the course in which participants were asked to review decisions made for particular scenarios. The importance of experience and the sharing of experience during on-the-job training were emphasized, as was the benefit of postevent review and discussion, given the name of *bumper talk*, a term commonly used in the service.

The intention of the course was to focus on changing behavior rather than providing a list of circumstances in which dangerous elements had already been identified. This was a departure from previous occupational safety and health programs that outlined the common risk situations, parameters to be alert for, and a cataloging of high-risk potential. Following the course, the same 12 scenarios and forced-choice response questionnaire were administered, together with the participant indication of what course of action they believed that the management would wish them to choose for each scenario.

The course was well received by the participants who felt that the content was relevant to practice (85% agree, strongly agree), was well presented (92% agree, strongly agree), and was potentially useful to paramedics (85% agree, strongly agree) (Klein and Donn, manuscript in preparation). There was a strong correlation between what the paramedics indicated they would do and what they perceived the organization would wish them to do.

- The precourse and postcourse data indicate that there were changes in selection of options more closely aligning with the perceived safe option identified as a management desire. Similarly, the shift was toward safer options from riskier. The paramedic responses to the 12 risk scenarios changed to selecting a less risky alternative after taking the course, significant at the 0.001 level.
- The paramedic responses to the 12 risk scenarios after taking the course, representing their view of what management would wish them to do, shifted to a less risky choice (significant at .001 level).
- There was no statistically significant difference in injury rate between the group who took the course, before and after, and those who did not take the course, where injury rate is the proxy measure for risk avoidance (Donn, 2011).

The sustained adoption of the principles of the SAFER course has been at least anecdotally reported, with specific applications to a biological hazard situation and a carbon monoxide scene. In both instances, the paramedics recalled the concepts of SAFER and did not take the seemingly appropriate, though risky choice of immediately entering the scene. In the latter instance, two paramedics who did enter the scene and required subsequent medical treatment had not taken the SAFER course. The organization has not kept specific records detailing the impact of the course.

The Organizational Commitment to Safety Survey revealed a wide range of views, with a diffuse distribution on the questions pertaining to organization role, and a narrower range on questions dealing with patient safety aspects or with paramedic safety. The strong sense of duty to the patient is evident in the results.

In this instance, what seemed more relevant in response to the loss of life tragedy was to provide the learners with a better understanding of the characteristics of decision makers in high-stress situations, and how the approach—not presented as prescriptive but rather descriptive of what decision makers do—may enable individuals to make better decisions. The emphasis was on three factors.

- First, SA is key to identifying the elements that pose risk.
- Second, the benefit of openness and attentiveness to cues depended partly on experience, partly on understanding of common ground, and partly on resistance to a natural tendency of emergency service workers to act first then analyze.
- The third element was that debriefing following a comprehensive approach allowed for the benefit of learning from each episode that presented risk elements, either familiar or new.

The sharing of experiences enables individuals new to the setting to benefit from the lessons, learned through reflection, by those who had more practice. This was elaborated in the bumper talk and "on-the-job training" components of the course. The development of a comprehensive approach to decision analysis followed the approach taken in other NDM-based courses, outlining the aspects of "what makes the decision difficult," what are the cues, and what are the options. Application of this approach in the course was less successful. Learners and instructors appeared more likely to consider the exercise as one in which the right answer was the important result, rather than acknowledging that the answer derived from analysis would be the right answer for the individual, and their benefit derived from considering what made the decision difficult for them, and then gaining new knowledge from what others considered.

The results of the initial data collection provide an interesting insight into the difficulties of obtaining behavioral change with learners who are more inclined to an action-oriented approach in their work setting. The reaction of learners was positive to the course content and the presentation approach, and the relevance of the curriculum to practice was clear. However, there was a minority who saw adoption of the approach as likely for their own practice (34% attitude change; 37% knowledge change; 47% improved decision-making; 40% awareness of safety issues increased).

The seeming mismatch of relevance but nonadoption can be better understood by examining the results of the organizational climate of a safety survey where a large percentage of respondents indicated that their practice was safe (86%). Other paramedics (82%) were not perceived as safe in their actions as the individual respondent. It is thus understandable that the numbers who indicate their practice will change are lower than anticipated. They do not see the need to change what they believe is a safe practice. Within social psychology, this illusory superiority is common. And, the work of Wyatt (2003) is consistent with this view.

Similarly, the variation in perception of the organizational commitment to safety is at odds with the dedication of individuals to patient safety. Thus, while they perceive that the organization is not as committed to safe practice in terms of either policy, management action, or provision of suitable equipment, the individual paramedic will operate in a safe manner for the protection of the patient despite these shortcomings. There is a consistency of this dichotomy with the analysis of paramedics from a folklore perspective (Tangherlini, 2000).

The tales that paramedics tell reflect a bravado and belief that the individual practitioner overcomes the mistakes of management and the underappreciation of their skills by physicians and other practitioners and is, at all times, working for the benefit of the patient (Tangherlini, 2000). While these attributes contain an element of truth, for the average paramedic the sense of being the frontline, unsupported provider of prehospital healthcare almost dictates placing personal safety after that of patients, partners, and others. The individualism of practice supports the oft-prevailing view that individuals commit errors or mistakes as a result of their own incompetence or weakness (Misasi et al., 2014).

Adoption of the principles of NDM may encounter obstacles other than those inherent in the traditional role as seen by paramedics. Paramedic practice has, in general, been protocol driven—follow specific steps that are constructed on the basis of fitting with most cases, doing no harm, and ensuring that scope-of-practice boundaries are not violated. Moving to another frame wherein there is greater individual responsibility for decision-making about patient treatment presents new challenges, relevant to the adoption of the RPD model for making safe decisions in the field.

The security that comes from having a prescribed direction is strong. Following a standard scene assessment approach may increase the SA but does not ensure the quality of decisions for action. Reflection and mental simulation of potential outcomes, recognition of changing risks, and the impact of new information or changes in the time sensitivity for treatment based on patient condition all have an effect on the treatment options available and chosen. The tendency of learners was to consider that the RPD model was in the mode of a protocol—do this, do this, and do this, and the decision is readily apparent. Rather than seeing the benefit as coming from reflection of cues, decisions, and adjustments to actions, a protocol-like sense continues (see for example Schmidt et al., 2000).

More deeply, there may be a need to consider the nature of learning approaches that the individuals bring to the education that is attempting to change behavior (Dreyfus, 2004). Here, the work of Marton and Säljö (1976a,b) informs the analysis. Their work on "surface" and "deep" learners, originally with students of economics and later (Marton and Booth, 1997) with university courses and students, in general, seems particularly relevant. Surface learners concentrate on the text, while deep learners focus on the meaning of the text. Those who, in later work, became described as "strategic" learners were able to adjust their learning mode to match with the expectations of the instructor.

How does this help to explain the apparent lack of impact of the SAFER course on practice? The distribution of learners between the surface and deep groups is not known for the paramedic population, but there is no evidence that would indicate it to be other than that found in the general postsecondary population. Thus, surface learners will have greater difficulty in adopting the reflective approach that underlies the RPD model since it requires constant adjustment, and the expectation of the learner is for a prescribed path—a protocol for decision-making. Since they already believe their practice to be safe, there is little incentive to adopt another view of decision-making informed by enhanced SA. Not only does it represent a significant change, it also requires that the meaning of the proposed model rather than the text of the model be understood.

For the deep learner, the meaning of the approach may be clearer. The inclusion of reflection in understanding and reviewing experience is not a major change. It is to this group that the approach presented in the SAFER course can be seen as relevant to their safe practice, as well as having application to the adoption of treatment guidelines for patient treatment. The premise of moving to treatment guidelines from protocols is that the paramedic at the patient side is in a better position to make the best treatment decision for that patient, rather than a group of physicians gathered at a table considering how to prescribe treatment that will apply in most cases, will have minimal harm potential, and will fit within the various scope-of-practice definitions for different license levels. Those who adopt treatment guidelines are treating the patient, rather than trying to make the patient fit a protocol.

Similarly, for a safe practice involving decision-making, the deep learner can recognize that the changing nature of situations means that it is impossible to define a "protocol of practice" for every situation. What is needed is adoption of an approach that will enable safe decisions to be made in a wide variety of situations—identify cues, mentally simulate results to a course of action, and adjust the actions accordingly.

It is only recently that the surface/deep distinction has appeared as a factor for consideration in EMS education (Heijne-Penninga et al., 2010). The authors structure their argument in terms of "need for cognition" related to an individual predilection to engage in effortful thinking. While students with a higher need for cognition performed better on both open- and closed-book tests, there was no influence of deep learning on performance. A potential extension of this finding would be to the nature of problems that paramedics face and how they are assessed. That is, do the assessments relate to the core knowledge of paramedic practice referred to by Heijne-Penninga et al. (2010) or the "nice-to-know" knowledge? As the role and scope of practice are expanding, the influence of deep or surface learning will become a more important aspect to consider in evaluation.

UPDATE ON DECISION-MAKING CONSIDERATIONS

Effective decision-making in the chaotic field of EMS practice can benefit from an understanding of the nature of decision-making and the specific aspects that are relevant to healthcare. Much of the work that has been done in this area focused on decision-making by physicians. In particular, the work of Croskerry (2002, 2006, 2007) is appropriate to guide applications in EMS. Klein's work (Ross, Klein et al., 2004; Klein, 2008, 2009) in establishing an understanding of the factors to be considered in NDM and the development of the RPD model provides a mechanism to assist practitioners. The important elements are that the model is descriptive rather than prescriptive, and creating the environment to acquire an understanding of the significant elements is the first step to making progress in this area.

The description of system I and system II as espoused by Kahneman (2011) is not in opposition to the RPD model (Kahneman and Klein, 2009), and both can be informative. The work of Croskerry (2007, 2009b,c) has investigated both the practice of decision-making in emergency medicine and the potential impact in education. The work of Norman and Eva (2010), in particular, points out the need to consider the ways of embedding these skills in the courses for medical students as well as how faculty can structure programs to give students the needed practice and opportunity of self-reflection in this regard. In particular, the use of simulation with well-structured debriefing sessions affords participants the experiences that will enable them to hone their insights.

It is perhaps useful to consider the realm of decision-making that has been reasonable for EMS practitioners as the field emerged. Rule-of-thumb decision approaches, based on informed medical considerations, provided an appropriate guideline for paramedics and often were the basis for the standardized protocols to be followed. The problem of clear thinking in an emergency (Schull et al., 2001) supports the benefit of heuristics to make the knowing of what is right to do translated more quickly into the doing. Mencl et al. (2013) raise some concerns regarding the ability of paramedics to correctly identify the ST elevation myocardial infarction (STEMI) presentations in their study. However, the limitations of evaluating the training programs including the assessment and reinforcement leave the question of suitability of the application of STEMI recognition by paramedics somewhat open.

A specific application for paramedics, START (Simple Triage and Rapid Treatment, Cook, 2001), was refined post-9/11 to improve the triaging of adult patients. And, perhaps, the exploration of a heuristic for detection of strokes, indicated as being more accurate than magnetic resonance imaging exams (Gigerenzer and Gaissmaier, 2011), suggests that specifically created heuristics appropriate to paramedic practice may represent an approach to enabling them to develop expertise relevant to their expanding practice. However, the insights of Feltovich et al. (2004) indicate caution.

Gigerenzer and Gaissmaier (2011) summarize the review of heuristics with indicators that can be applied to EMS activities. Heuristics are presented as potentially more accurate than complex strategies, as having a context dependency for accuracy and being suitable to function as one of an available array of choices for decision-making. These aspects support the use of heuristics in EMS practice thus far.

Decision-making in EMS practice has not received the study that decision-making by physicians has. The exploration of decision-making in emergency medicine, for example by Croskerry (2002, 2008, 2009a,b,c) can be placed into three large areas—first, the inherent factors which interfere with decision-making; second, the consideration of a theory to better understand the decision-making process and its relationship to expertise; and, third, what aspects of decision-making can be taught.

Croskerry (2002) explores the reasons that understanding how physicians think in the ED setting is a significant research area. The use of heuristics is acknowledged, and the benefit of being more aware of cognitive processes is seen as informing the quality of decisions (e.g., Flavell, 1976), improving instruction in the teaching of cognitive strategies, and, most importantly, improving patient safety. The discussion of the types of biases that influence decision-making has relevance to the EMS setting.

A full discussion of bias and its influence on clinical decision-making was reported by Croskerry (2002). The use of specific cognitive strategies to mitigate the identified biases can be applied to both medical education and professional development. Croskerry (2002) refers to the importance of understanding the role of metacognitive strategies as a characteristic of experts who have been successful in "overcoming weak cognitive tendencies, biases, and flawed heuristics." His work was focused on decision-making within the ED and thus seems of relevance to EMS situations. In particular, there may be associations that are more prevalent for low-volume EMS stations where particular patient presentations are seen so infrequently that incorrect application of heuristic thinking would be understandable. However, it should be noted that a number of decision researchers have been critical of this bias and flawed heuristic view (see Klein, 2009, for a summary). Biases can be demonstrated in laboratory settings but miss the value of heuristics for operating in complex environments.

Other works based on physician decision-making explored which theoretical model best matched decisions made under diagnostic uncertainty. The dual-processing model was most closely aligned with the physician decision-making studied, in particular, a threshold approach in which the decision to treat was applied when the probability of disease was above a predetermined threshold and the decision not to treat when the presentation was below the threshold. Developing such an approach may be of benefit to paramedic practice (Djulbegovic et al., 2014).

The analysis of decision-making also introduces the concept of errors in decision-making, in the realm of clinical decision-making by physicians (Shaban, 2005; Croskerry, 2006). Again, the identification of the cognitive basis for the errors raises the prospect that there can also be approaches to minimizing or mitigating the cognitive biases. Another view of error in clinical decision-making is presented by Thompson and Dowding (2004), who identify several sources of error. Strict adherence to rules categorized as "bad," false, or inadvisable is one source. The inappropriate application of heuristics is seen as another bias-laden issue. The authors present approaches to overcoming these error sources, and the extension beyond the realm of nursing practice may be worth considering for other health practitioners.

A further category as a contributory source to error in clinical decision-making is overconfidence. This arena has been reviewed by several authors (Kruger and Dunning, 1999; Davis et al., 2006; Berner and Graber, 2008; Croskerry and Norman, 2008), and there are consistencies in the findings over the number of studies. Reliance on the initial consideration of the situation, application of inappropriate heuristics, and the inability to recognize one's own limitations are presented as the major sources of difficulty. There are varying degrees to which successful interventions may be applied in each instance. Understanding the role of cognitive strategies is presented as one way in which practitioners can reduce the errors to improve patient safety (Chapman and Aubin, 2002). The comment by Berner and Graber regarding physician error, "that they believe that their initial diagnoses are correct (even when they are not) and there is no reason for change," is reminiscent of the intervention previously described to improve paramedic SA. They perceived their practice to be safe and hence were not as receptive to education designed to improve that aspect of decision-making. The review by Davis et al. (2006) suggests that physicians are not proficient at accurate self-assessment. This inability resulted in those least skilled having the worst self-assessment and the greatest overconfidence.

Improvements may be assisted by reviewing the aspects of critical thinking as outlined by Facione and Facione (2008). Their summary points to recent trends to improve clinical judgment training, as well as the apparent improvements in outcomes with an increased application of case-based and problem-based learning in educational programs in healthcare, for example Critical Care Paramedic education (BC Ambulance Service, 2013). This opens consideration of how decisions are being made in clinical settings (Sullivan and Chumbley, 2010). In general, the dual-processing model (system I/system II) as explored by Sanfey and Chang (2008) and Kahneman and Klein (2009) reinforces the complexity of decision-making in the NDM environment where reliance solely on System I (or the recognition-primed part of RPD) while efficient has potential risk for error. The more effortful system II reflection can be essential to achieving the status of expert in a particular field, as alluded to previously (Croskerry, 2002; Frederick, 2005; Croskerry and Norman, 2008). The importance of a full understanding of the context in which a decision is made must be emphasized, if there is to be learning from decision errors, whether system I or system II based (Croskerry, 2009c).

The broad area of education in decision-making, an approach to expertise acquisition, incorporates several strategies (e.g., Bowen, 2006; Milkman et al., 2008; Okoli et al., 2013). The general trend is to examine the approaches of experts and make them available to novices. The requirement is for the expert to assess both the nature of the problem presented and the developmental stage of the novice to ensure an appropriate level of feedback and probing questions. Concomitant considerations are of the roles of system I and system II thinking, the impact of prior experience as cues in RPD, and whether distinct approaches to unravel these two roles, the immediate and the reflective, in improving decision-making can be easily resolved (Milkman et al., 2008). Assistance in developing suitable approaches may be found in the work of Evans (2008), which takes a broad view of the available literature on dual processing in higher cognition. Further, there is an extension of these concepts to include categorization of decision-making processes, leading to the incorporation of intuitive judgment and intuition as a reflection of expertise (Glöckner and Witteman, 2010; Chu and MacGregor, 2011; Klein, 2013). Cunningham et al. (2009) in detailing a categorization of insights offer some potentially fruitful concepts for further consideration of the topic.

The approaches to improving clinical decision-making by use of cognitive strategies have been explored for emergency medicine and provide a guideline for improving the expertise in EMS (Chapman and Aubin, 2002). But as pointed out by Lajoie (2003), the approach to transition from novice to expert requires first that some form of assessment of what constitutes expertise for a particular domain is necessary. Novices can then benefit from awareness of the intended path to improve their expertise. Smith et al. (2013) engaged in such a cognitive task analysis to compare the performance of experienced and less experienced paramedics in their approach to complex patient presentations. While their findings demonstrated the greater expertise of more experienced practitioners and their greater gathering of cues, use of inferential reasoning, and strategic thinking, the shortage of research specific to EMS is alluded to in their proposals for future research in this area. The use of cognitive forcing strategies to reduce biases has been ineffective in at least one study (Sherbino et al., 2014) and thus reinforces the need for additional study in the area of increasing expertise. The poor track record of debiasing efforts calls into question the efficacy of the cognitive bias approach.

In a related healthcare field, nursing, other approaches have been introduced to promote critical thinking, a valuable line of inquiry to improve decision-making (Vacek, 2009). The relevance to EMS expertise is likely to be one of accepting relevant research from studies of other health professions. Indeed, the outlines of research on emergency physicians including their decision-making, cognitive strategies to mitigate bias, and ways of applying lessons learned to the education of medical students can be a guiding framework for acquisition of expertise in EMS practice. The available research on EMS paramedics is only recently concerned with these areas of inquiry.

Jensen et al. (2011) and Jensen (2011) present investigations of paramedic decision-making in high-acuity calls and outline the strategies employed for thinking about the patient presentation and the most appropriate treatment. The unique contexts in which paramedic practice occurs have features of high-density decision-making during the initial on-scene phase, and expertise in that portion of patient treatment differs markedly from a potentially more controlled setting for other emergency health practitioners. The work of Ryan and Halliwell (2012) outlines two approaches that are deemed involved in paramedic decision-making and expertise and their relationship to deep and surface learning strategies. Earlier studies (Heijne-Penninga et al., 2010) explored the impact of deep learning on effective preparation for medical student assessments.

There appear to be several approaches that can have an impact on the improvement of clinical decision-making and expertise within the field of medicine. The adaptation of these specifically to paramedic expertise may seem beneficial to improving practice, but a cautionary view is presented by Graber (2009) with the question of whether educational strategies to reduce error can, in fact, be taught. Eva (2005) provides guidelines for clinical instructors regarding what they need to know about clinical reasoning, for effective teaching. Norman and Eva (2010) and Norman (2005, 2009) review research in the field of clinical reasoning and the relationship of dual processing in understanding diagnostic errors. Again, the research from the field of physician education and practice is viewed as a guideline for implementation in improving EMS expertise.

RECENT WORK ON THE EXPERIENTIAL/RATIONAL BASIS OF PARAMEDICS IN DECISION-MAKING

A further area of research has been the relationship between the experiential and rational bases for decision-making as it relates to tolerance for ambiguity. In this area, there appear to be strong correlations for cardiologists, emergency physicians, and paramedics (Calder et al., 2012; Jensen et al., 2014a,b). The demonstration of expertise in these contexts may be posited to reflect the ability to function effectively under the conditions of NDM, the frequent context for EMS practice (Furnham and Marks, 2013). This approach is based more on the individual capability rather than knowledge or skills resulting from education. For paramedics to benefit from the work on decision-making, the question is what approaches would benefit those who see their decision-making as having a rational basis rather than acknowledging the experiential basis. Individuals attracted to the profession may first benefit from an understanding of their decision-making (Klein, 1998, 2003, 2008) as well as becoming more self-reflective (Marton and Booth, 1997; Kruger and Dunning, 1998). Other work (Croskerry et al., 2010) introduces the element of emotional influences in clinical decision-making. For paramedics, this further understanding of aspects of their decision-making beyond the rational analysis can be a significant contribution to improved decisions. Williams et al. (2013) likewise point out that changing attitudes will be a necessary component if knowledge translation is to achieve a more influential position in adjusting paramedic practice to the new demands of the profession.

There may be value in approaching the educational components required for acquisition of expertise in the new environment by consideration of more fundamental aspects—the philosophical distinction of "knowing how" and "knowing that" (Quay, 2004), for example. Other potential sources for consideration would be earlier discussions of cognitive study to the understanding of expertise (Schmidt et al., 1990) or exploration of applications of the reflection (system II) on initial decisions (system I) in terms of "slowing down," a counterintuitive notion for traditional paramedic reactive approaches to treatment (Moulton et al., 2007; Di Stefano et al., 2014).

FUTURE RESEARCH

At least 3 areas for future research seem relevant to the changing role of paramedics. The expanding scope of practice in terms of a larger response to nonurgent, nonemergent patient presentations will require an understanding and, perhaps, greater concentration on nontechnical skills. These attributes have been the subject of research and study in other medical areas, for example, surgery and anesthesiology (Flin and Maran, 2004; Flin et al., 2007), and attention to the needs for paramedics in this realm seems reasonable.

With paramedics taking a more proactive rather than reactive role, community paramedicine as an example, research on the effectiveness of such roles will be important to best tailor the role (Bigham et al., 2013) and the interprofessional relationships (Misasi et al., 2014) that are required to ensure the wise investment of healthcare resources. In this specific area, two approaches bear watching. The work of Morrison (Pers. Comm; Drennan et al., 2014) investigating, in a systematic fashion, the impact of community paramedicine practice on three chronic conditions, congestive heart failure (CHF), diabetes, and chronic obstructive pulmonary disease (COPD), will provide insights into anticipated effectiveness of this expanded role. Another view is taken by Craig (2015a,b), who suggests that confirmatory research is needed to make the ready acceptance of this trend valid.

A second broad area for future research would be on the role of technology, both in terms of paramedic practice and in regard to the educational approaches for the new roles, and the scope of current practice. Consider first the impact of technology in new roles. The impact of telehealth applications, for example, to enable more rapid and continuous contact with medical oversight may permit an increase in scope of practice to facilitate the realization of the intent to take the ED to the field. There have already been some approaches to facilitating such applications (EMS World, 2016). The confirmation of expenditures in technology as a means to provide more cost-effective healthcare seems reasonable as the direct costs of hospitalization are already identified (British Columbia Ministry of Health, 2014).

Further to the increasing use of technology are field applications for items such as intubation assists and cardiopulmonary resuscitation (CPR) assists. While the evidence is mixed for the use of intubation in the field (Wang et al., 2015), the role of CPR assists has some potential benefit (Krep et al., 2007). There is not a clear role for such assists in terms of patient benefit, but there may be in terms of paramedic safety to provide effective CPR while in transport.

For these and other potential interventions, large-scale randomized controlled trials, although expensive to conduct, are necessary to provide the evidence of evidence-based medicine. Paramedic involvement in research (Mencl et al., 2013; Mausz and Cheskes, 2015) is a two-edged sword. On one hand, the data collection from the field is essential. On the other, the potential impact in terms of self-initiated practice without the clear evidence is consistent with the nature of paramedic practitioners (Tangherlini, 2000).

Technology can also play a role in the education of paramedics for their current or new roles. In the current scope of practice, simulations may be provided for initial acquisition of skills as well as the maintenance of competency in those skills. This area is one which, again, has roots in established medical interventions (Cheng et al., 2014; Halliwell et al., 2015) and is increasingly found in paramedic research. The use of simulation has been widely accepted in medical education and can be used to enhance new skill sets, to maintain existing skills, or to assess the effectiveness of such interventions in clinical practice (Tavares et al., 2013). Some aspects of the expanded role in terms of the nontechnical skill acquisition may be influenced by applications such as ShadowBox (Klein et al., 2015), which have already been demonstrated as effective in the realm of "Good Strangers," a program on cultural competency for the US Army. In this application, there was demonstrated success in providing individual learning through computer-based, virtual and simulation activities. These early demonstrations of improved educational outcomes suggest that further technology applications may be of benefit. However, there would appear to be a required understanding of decision-making and technology applications to ensure success (Patterson et al., 2013) as well as a thorough assessment of the knowledge of the "experts" in the domain (Hoffman et al., 1998).

Virtual reality applications such as TheraSim have been used in the provision of education for healthcare providers in a number of underserved areas, and similar approaches may be cost-effective methods for remote and rural areas where paramedic practice is provided frequently by individuals who have the least training and the least experience and, in low-call volume areas, the least opportunity to consolidate, improve, and maintain their skills. The need for research to provide evidence and direction for such applications is a ripe area for consideration.

The technology impact on practice can be determined only by costly large-scale, randomized controlled studies. But small-scale research will continue to provide guidance for adjusting and enhancing paramedic response to the new needs. Studies involving technology that can impact practice are suggested by an exploration of paramedic safe practice in a common area of service—cardiac arrest resuscitation (Ross et al., 2015). In this instance, eye-tracking technology revealed an inattention to individual and partner safety during defibrillation. The "presumed expertise" within even a frequently practiced intervention underlines the need for reinforcement of best practice for continued retention of expertise in a complex skill.

Overarching these potential research applications is the common issue of the demographics of the paramedic practitioners. Individuals with extensive experience will be taking the corporate memory of paramedic practice with them, and application of some methods to preserve the valuable insights gained over years seems indicated. Eliciting expertise is a resource-intensive field that can lead to an accelerated acquisition of expertise, as discussed thoroughly by Hoffman et al. (2014). Specific suggestions for accelerating expertise (Fadde and Klein, 2012; Fadde, 2013) offer some insights to achieve this desirable outcome. In light of the increasing complexity of patient treatments being offered in the field by paramedics, the ability to accelerate their path to expert is desirable. However, as O'Neil and Addrizzo-Harris (2009) point out, the effectiveness of continuing education to improve skills and knowledge can be enhanced by multiple exposures and longer duration of education sessions. These recommended practices may be at odds with current approaches to continuing education for paramedics, either for license or certification retention or as provided by the employer.

CONCLUSION

The aspect of demonstrating expertise for paramedics is challenged in a number of ways. The research evidence specific to EMS that would assist in improving decision-making is scant. Only recently has there been attention to the notions of clinical decision-making in EMS practice, which is only recently being recognized as a distinct field of practice in healthcare.

The evolution of the role is expanding both in allowing treatment to more complex patients in the prehospital setting and in providing care in a community role as an adjunct to primary care providers. These two developments are almost in conflict—the former requiring more specialized skills and knowledge and the latter demanding more generalized knowledge with a greater emphasis perhaps on the nontechnical skills. And within each area, the role of technology places added assistance such as telehealth while simultaneously increasing demands for greater technical knowledge. The facility with technology may prove less of a barrier for more recent hires but can leave the experienced practitioner at a disadvantage.

Finally, achieving the path to professionalism has its own conflicts in terms of expertise in practice and expertise in theoretical knowledge. Working in the interprofessional setting, more likely with the expanded community role, will require changes to the education approach for the new generation of paramedics with expertise.

REFERENCES

Andrew, A., de Wit, A., Meadley, B., Cox, S., Bernard, S. & Smith, K. (2015). Characteristics of patients transported by a paramedic-staffed helicopter emergency medical service in Victoria, Australia. *Prehospital Emergency Care, 19*, 416–424.

Atack, L. & Maher, J. (2010). Emergency medical and health providers' perceptions of key issues in prehospital patient safety. *Prehospital Emergency Care, 14*, 95–102.

BC Ambulance Service. (2013). Critical Care Paramedic Education Program: Module 103: Critical thinking and clinical decision-making.

Berner, E.S. & Graber, M.L. (2008). Overconfidence as a cause of diagnostic error in medicine. *American Journal of Medicine, 121*, S2–S23.

Bigham B.L., Kennedy, S.M., Drennan, I. & Morrison, L.J. (2013). Expanding paramedic scope of practice in the community: A systematic review of the literature. *Prehospital Emergency Care, 17*, 361–372.

Bigham, B. & Welsford, M. (2015). Applying hospital evidence to paramedicine: Issues of indirectness, validity and knowledge translation. *Canadian Journal of Emergency Medicine, 17*, 281–285.

Bowen, J.L. (2006). Educational strategies to promote clinical diagnostic reasoning. *New England Journal of Medicine, 355*, 2217–2225.

British Columbia Ministry of Health. (2014). Setting Priorities for the BC Health System. http://www2.gov.bc.ca/gov/content/health/about-bc-s-health-care-system/health-priorities /setting-priorities-for-bc-health

Calder, L.A., Forster, A.J., Stiell, I.G., Carr, L.K., Brehaut, J.C., Perry, J.J. et al. (2012). Experiential and rational decision making: A survey to determine how emergency physicians make clinical decisions. *Emergency Medicine Journal*, *29*, 811–816.

Chapman, D.M. & Aubin, C.D. (2002). Cognitive strategies for reducing errors in clinical decision making. SAEM Didactic Session. St. Louis, MO.

Cheng, A., Auerbach, M., Hunt, E.A., Chang, T.P., Pusic, M., Nadkarni, V. et al. (2014). Designing and conducting simulation-based research. *Pediatrics*, *133*, 1081–1101.

Chu, Y. & MacGregor, J.N. (2011). Human performance on insight problem solving: A review. *Journal of Problem Solving*, *3*, 119–150.

Cook, L. (2001). The World Trade Center attack—The paramedic response: An insider's view. *Critical Care*, *5*, 301–303.

Craig, A. (2015a). *New technology is changing how we provide clinical care*. Retrieved from http://www.jems.com/content/jems/en/articles/supplements/special-topics/on-the-leading-edge/new-technology-is-changing-how-we-provide-clinical-care.html

Craig, A. (2015b). *AMR VP of clinical strategies talks mobile integrated health care*. Retrieved from http://www.ems1.com/columnists/paramedic-chief-column/articles/1977808-AMR-VP-of-clinical-strategies-talks-mobile-integrated-health-care/

Croskerry, P.A. (2002). Achieving quality in clinical decision making: Cognitive strategies and detection of bias. *Academic Emergency Medicine*, *9*, 1184–1204.

Croskerry, P.A. (2006). Critical thinking and decision making: Avoiding the perils of thin-slicing. *Annals of Emergency Medicine*, *48*, 720–722.

Croskerry, P.A. (2007). The cognitive imperative: Thinking about how we think. *Academic Emergency Medicine*, *7*, 1223–1231.

Croskerry, P.A. (2009a). Context is everything or how could I have been that stupid? *Healthcare Quarterly*, *12*, 171–177.

Croskerry, P.A. (2009b). Universal model of diagnostic reasoning. *Academic Medicine*, *84*, 1022–1028.

Croskerry, P.A. (2009c). Clinical cognition and diagnostic error: Applications of a dual process model of reasoning. *Advances in Health Science Education*, *14*, 27–35.

Croskerry, P. & Norman, G. (2008). Overconfidence in clinical decision making. *American Journal of Medicine*, *121*, 24–29.

Croskerry, P., Abbass, A. & Wu, A.W. (2010). Emotional influences in patient safety. *Journal of Patient Safety*, *6*, 199–205.

Cunningham, J.B., MacGregor, J.N., Gibb, J. & Haar, J. (2009). Categories of insight and their correlates: An exploration of relationships among classic-type insight problems, rebus puzzles, remote associates and esoteric analogies. *Journal of Creative Behaviour*, *43*, 262–280.

Davis, D.A., Mazmanian, P.E., Fordis, M., Van Harrison, R., Thorpe, K.E. & Perrier, L. (2006). Accuracy of physician self-assessment compared with observed measures of competence. *Journal of American Medical Association*, *206*, 1094–1102.

Di Stefano, G., Gino, F., Pisano, G. & Staats, B. (2014). *Learning by thinking: How reflection aids performance*. Harvard Business School Working Paper, 1–48. http://www.sc.edu/uscconnect/doc/Learning%20by%20Thinking,%20How%20Reflection%20Aids%20Performance.pdf

Djulbegovic, B., Elqayam, S., Reljic, T., Hozo, I., Miladinovic, B., Tsalatsanis, A. et al. (2014). How do physicians decide to treat: An empirical evaluation of the threshold model. *Biomed Central: Medical Informatics and Decision Making*, *14*, 1–10.

Donn, J.S. (2011). Can an RPD model based course change behaviour? One experience with paramedics. *Proceedings of the 10th International Naturalistic Decision Making Conference*, Orlando, FL.

Drennan, I.R., Dainty, K.N., Hoogeveen, P., Atzema, C.L., Barrette, N., Hawker, G. et al. (2014). Expanding paramedicine in the community (EPIC): Study protocol for a randomized controlled trial. *Trials*, *15*, 473.

Dreyfus, S.E. (2004). The five-stage model of adult skill acquisition. *Bulletin of Science, Technology & Society*, *24*, 177–181.

EMS World. (2016). *Telemedicine today*. Retrieved from http://www.emsworld.com/article /12184039/telemedicine-today-part-1-getting-started

Endsley, M.R. (1995). Toward a theory of situation awareness in dynamic systems. *Human Factors*, *37*, 32–64.

Ericsson, K.A. (1996). The acquisition of expert performance: An introduction to some of the issues. In K.A. Ericsson (Ed.), *The road to excellence: The acquisition of expert performance in the arts and sciences, sports, and games* (pp. 1–50). Mahwah, NJ: Erlbaum.

Eva, K.W. (2005). What every teacher needs to know about clinical reasoning. *Medical Education*, *39*, 98–106.

Evans, J.S. (2008). Dual-processing accounts of reasoning, judgment and social cognition. *Annual Review of Psychology*, *59*, 255–78.

Facione N.C. & Facione P.A. (2008). *Critical thinking and clinical reasoning in the health sciences: An interdisciplinary teaching anthology*. Millbrae, CA: California Academic Press.

Fadde, P. & Klein, G. (2012). Accelerating expertise using action learning activities. *Cognitive Technology*, *17*, 11–18.

Fadde, P. (2013). Accelerating the acquisition of intuitive decision-making through expertise-based training (XBT). *Paper presented at the meeting of Interservice/Industry Training, Simulation and Education Conference (I/ITSEC)*, Orlando, FL.

Feltovich, P.J., Hoffman, R.R., Woods, D. & Roesler, A. (2004). Keeping it too simple: How the reductive tendency affects cognitive engineering. *Institute of Electrical and Electronics Engineers: Intelligent Systems*, *19*, 90–94.

Flavell, J.H. (1976). Metacognition and cognitive monitoring: A new area of psychological inquiry. *American Psychologist*, *34*, 906–911.

Flin, R. & Maran, N. (2004). Identifying and training non-technical skills for teams in acute medicine. *Quality & Safety in Health Care*, *13*, 80–84.

Flin, R., Yule, S., Paterson-Brown, S., Maran, N., Rowley, D. & Youngson, G. (2007). Teaching surgeons about non-technical skills. *Surgeon*, *5*, 86–89.

Frederick, S. (2005). Cognitive reflection and decision making. *Journal of Economic Perspectives*, *19*, 25–42.

Furnham, A. & Marks, J. (2013). Tolerance of ambiguity: A review of the recent literature. *Psychology*, *4*, 717–728.

Gigerenzer, G. & Gaissmaier, W. (2011). Heuristic decision making. *Annual Review of Psychology*, *62*, 451–482.

Glöckner, A. & Witteman, C. (2010). Beyond dual-process models: A categorisation of processes underlying intuitive judgement and decision making. *Thinking and Reasoning*, *16*, 1–25.

Graber, M.L. (2009). Educational strategies to reduce diagnostic error: Can you teach this stuff? *Advances in Health Sciences Education: Theory and Practice*, *14*, 63–69.

Hamid, H.A.S., Waterson, P. & Hignett, S. (2009). Situation awareness to support decision-making among emergency care practitioners. *Proceedings of NDM 9, the 9th International Conference on Naturalistic Decision Making, UK*.

Halliwell, D., Ryan, L. & Jones, P. (2015). Designing prehospital medical simulation scenarios. http://healthysimulation.com/2858/designing-prehospital-medical-simulation-scenarios/

Heijne-Penninga, M., Kuks, J.B.M., Hofman, W.H.A. & Cohen-Schotanus, J. (2010). Influences of deep learning, need for cognition and preparation time on open- and closed-book test performance. *Medical Education*, *44*, 884–891.

Hickson, H., Williams, B. & O'Meara, P. (2015). Paramedicine students' perception of preparedness for clinical placement in Australia and New Zealand. *BioMed Central: Medical Education*, *15*, 168.

Hobgood, C., Xie, J., Weiner, B. & Hooker, J. (2004). Error identification, disclosure and reporting: Practice patterns of three emergency medicine provider types. *Academy of Emergency Medicine*, *11*, 196–199.

Hobgood, C., Bowen, J.B., Brice, J.H., Overby, B. & Tamayo-Sarver, J.H. (2006). Do EMS personnel identify, report, and disclose medical errors? *Prehospital Emergency Care*, *10*, 21–27.

Hoffman, R.R., Crandall, B. & Shadbolt, N. (1998). Use of the critical decision method to elicit expert knowledge: A case study in the methodology of cognitive task analysis. *Human Factors*, *40*, 254–276.

Hoffman, R.R., Ward, P., Feltovich, P.J., DiBello, L., Fiore, S.M. & Andrews, D.H. (2014). *Accelerated expertise: Training for high proficiency in a complex world.* New York: Taylor & Francis.

Jensen, J., Tavares, W., Calder, L., Bienkowski, A., Walker, M., Travers, A. et al. (2014a). Experiential and rational clinical decision-making: A survey to determine decision-making styles of paramedic students. *Canadian Journal of Emergency Medicine*, *18*, 1–10.

Jensen, J.L., Travers, W., Calder, L., Bienkowski, A., Walker, M., Travers, A. et al. (2014b). Experiential and rational clinical decision-making: A survey to determine decision-making styles of paramedics. *Patient Education and Counseling*, *18*, 152.

Jensen, J.L., Croskerry, P. & Travers, A.H. (2011). Consensus on paramedic clinical decisions during high-acuity emergency calls: Results of a Canadian Delphi study. *Canadian Journal of Emergency Medicine*, *13*, 310–318.

Jensen, J. (2011). Paramedic clinical decision-making: Results of two Canadian studies. *International Journal of Paramedic Practice*, *1*, 186–194.

Joyce, C.M., Wainer, J., Piterman, L., Wyatt, A. & Archer, F. (2009). Trends in the paramedic workforce: A profession in transition. *Australian Health Review*, *33*, 533–540.

Kahneman, D. & Klein, G. (2009). Conditions for intuitive expertise: A failure to disagree. *American Psychologist*, *64*, 515–526.

Kahneman, D.K. (2011). *Thinking fast and slow.* Richmond, Canada: Farrar, Straus and Giroux.

Klein, G. (1998). *Sources of power: How people make decisions.* Cambridge, MA: MIT Press.

Klein, G. (2003). *The power of intuition.* New York: Crown Business.

Klein, G. (2008). Naturalistic decision making. *Human Factors*, *50*, 456–460.

Klein, G. (2009). *Streetlights and shadows: Searching for the keys to adaptive decision making.* Cambridge, MA: MIT Press.

Klein, G. (2013). *Seeing what others don't.* Philadelphia, PA: Public Affairs.

Klein, G., Borders, J., Wright, C. & Newsome, E. (2015). An empirical evaluation of the ShadowBox training method. *Proceedings of the 13th International Conference on Naturalistic Decision Making (NDM 2015)*, US, 9–12.

Klein, G.A., Orasanu, J. & Calderwood, R. (1993). *Decision making in Action: Models and Methods (Cognition and Literacy).* Praeger.

Krep, H., Mamier, M., Breil, M., Heister, U., Fischer, M. & Hoeft, A. (2007). Out-of-hospital cardiopulmonary resuscitation with the AutoPulse™ system: A prospective observational study with a new load-distributing band chest compression device. *Resuscitation*, *73*, 86–95.

Kruger, J. & Dunning, D. (1998). Unskilled and unaware of it: How difficulties in recognizing one's own incompetence lead to inflated self-assessments. *Journal of Personality and Social Psychology*, *77*, 1121–1134.

Lajoie, S.P. (2003). Transitions and trajectories for studies of expertise. *Educational Researcher*, *32*, 21–25.

Marton, F. & Booth, S. (1997). *Learning and awareness.* Mahwah, NJ: Lawrence Erlbaum Association.

Marton, F. & Säljö, R. (1976a). On qualitative differences in learning I: Outcome and process. *British Journal of Educational Psychology*, *46*, 4–11.

Marton, F. & Säljö, R. (1976b). On qualitative differences in learning II: Outcome as a function of the learner's conception of the task. *British Journal of Educational Psychology*, *46*, 115–127.

Mausz, J. & Cheskes, S. (2015). The impact of prehospital resuscitation research on in-hospital care. *Canadian Journal of Emergency Medicine, 17*, 551–557.

Mencl, F., Wilber, S., Frey, J., Zalewski, J., Maiers, J.F. & Bhalla, M.C. (2013). Paramedic ability to recognize ST-segment elevation myocardial infarction on prehospital electro-cardiograms. *Prehospital Emergency Care, 17*, 203–210.

Milkman, K.L., Chugh S. & Bazerman, M.H. (2008). How can decision making be improved? *Perspectives of Psychological Science, 4*, 379–383.

Misasi, P., Lazzara, E.H. & Keebler, J.R. (2014). Understanding multi-team systems in emergency care, one case at a time. In E. Salas, R. Rico & M. Shuffler (Eds.), *Pushing the boundaries: Multi-team systems in research and practice*. Bradford, UK: Emerald Group Publishing.

Moulton, C.E., Regehr, G., Mylopoulos, M. & MacRae, H.M. (2007). Slowing down when you should: A new model of expert judgement. *Academic Medicine, 82*, 109–116.

National Academy of Sciences. (1966). *Accidental death and disability: The neglected disease of modern society*. Washington, DC: National Academies Press.

Norman, G. (2005). Research in clinical reasoning: Past history and current trends. *Medical Education, 39*, 418–427.

Norman, G. (2009). Dual processing and diagnostic errors. *Advances in Health Science Education: Theory and Practice, 14*, 37–49.

Norman, G.R. & Eva, K.W. (2010). Diagnostic error and clinical reasoning. *Medical Education, 44*, 94–100.

O'Hara, R., Johnson, M., Siriwardena, A.N., Weyman, A., Turner, J., Shaw, D. et al. (2015). A qualitative study of systemic influences on paramedic decision making: Care transitions and patient safety. *Journal of Health Services Research and Policy, 20*, 45–53.

Okoli, J.O., Weller, G., Watt, J. & Wong, B.L.W. (2013). Decision making strategies used by experts and the potential for training intuitive skills: A preliminary study. *Proceedings of the 11th International Conference on Naturalistic Decision Making*, France.

O'Meara, P. (2009). Paramedics marching towards professionalism. *Australasian Journal of Paramedicine, 7*, 1–5.

O'Meara, P. (2011). So how can we frame our identity? *Journal of Paramedic Practice, 3*, 57.

O'Meara, P. & Grbich, C.F. (2009). *Paramedics in Australia: Contemporary challenges of practice*. Frenchs Forest, Australia: Pearson Education.

O'Meara, P.F., Williams, B. & Hickson, H. (2015). Paramedic instructor perspectives on the quality of clinical and field placements for university education paramedicine students. *Nurse Education Today, 35*, 1080–1084.

O'Neil, K.M. & Addrizzo-Harris, D.J. (2009). Continuing medical education effect on physician knowledge application and psychomotor skills. *Chest, 135*, 37–41.

Patterson, M., Militello, L.G., Taylor, R., Bunger, A., Wheeler, D., Klein, G. et al. (2013). Acceleration to expertise in healthcare: Leveraging the critical decision method and simulation-based training. *Proceedings of the 11th International Conference on Naturalistic Decision Making (NDM 2013)*, Marseille, France.

Phua, D.H. & Tan, N.C.K. (2013). Cognitive aspects of diagnostic errors. *Annals Academy of Medicine, Singapore, 42*, 33–41.

Quay, J. (2004). *Knowing how and knowing that: A tale of two ontologies*. Retrieved from https://staff.education.unimelb.edu.au/~jquay/QuayIOE2004.pdf

Ross, K.G., Klein, G.A., Thunhom, P., Schmitt, J.F. & Baxter, H.C. (2004). The recognition-primed decision model. *Military Review*, 6–10.

Ross, L., Williams, B. & Boyle, M. (2015). Defibrillation safety: An examination of paramedic perceptions using eye-tracking technology. *Association for Simulated Practice in Healthcare, 1*, 62–66.

Ryan, L. & Halliwell, D. (2012). Paramedic decision-making: How is it done? *Journal of Paramedic Practice, 4*, 343–351.

Sanfey, A.G. & Chang, L.J. (2008). *Of two minds when making a decision.* Scientific American: Mind Matters. June 3. Retrieved from http://www.scientificamerican.com /article.cfm?id=of-two-minds-when-making&print=true

Schmidt, H.G., Norman, G.R. & Boshuizen, H.P.A. (1990). A cognitive perspective on medical expertise: Theory and implications. *Academic Emergency Medicine, 65,* 611–621.

Schmidt, T., Atcheson, R., Federiuk, C., Mann, N.C., Pinney, T., Fuller, D. et al. (2000). Evaluation of protocols allowing emergency medical technicians to determine need for treatment and transport. *Academic Emergency Medicine, 7,* 663–669.

Schull, M.J., Ferris, L.E., Tu, J.V., Hux, J.E. & Redelmeier, D.A. (2001). Problems for clinical judgement: 3. Thinking clearly in an emergency. *Canadian Medical Association Journal, 164,* 1170–1175.

Shaban, R.Z. (2005). Theories of clinical judgment and decision-making: A review of the theoretical literature. *Journal of Emergency Primary Health Care, 3,* 23–33.

Sherbino, J., Kulasegaram, K., Howey, E. & Norman, G. (2014). Ineffectiveness of cognitive forcing strategies to reduce biases in diagnostic reasoning: A controlled trial. *Canadian Journal of Emergency Medicine, 16,* 34–40.

Smith, M.W., Bentley, M.A., Fernandez, A.R., Gibson, G., Shweikhart, S.B. & Woods, D.W. (2013). Performance of experienced versus less experienced paramedics in managing challenging scenarios: A cognitive task analysis study. *Annals of Emergency Medicine, 62,* 367–379.

Sullivan, D.L. & Chumbley, C. (2010). Critical thinking: A new approach to patient care. *Journal of Emergency Medicine, 35,* 48–53.

Tangherlini, T.R. (2000). Heroes and lies: Storytelling tactics among paramedics. *Folklore, 111,* 43–66.

Tavares, W., Leblanc, V.R., Mausz, J., Sun, V. & Eva, K.W. (2013). Simulation-based assessment of paramedics and performance in real clinical contexts. *Prehospital Emergency Care, 18,* 116–122.

Thomas, S.H., Brown, K.M., Olvei, Z.J., Spalte, D.W., Lawner, B.J., Sahni, R. et al. (2013). An evidence-based guideline for the air medical transportation of prehospital trauma patients. *Prehospital Emergency Care, 18,* 35–44.

Thompson, C. & Dowding, D. (2004). Awareness and prevention of error in clinical decision-making. *Nursing Times, 23,* 40–43.

Vacek, J.E. (2009). Using a conceptual approach with concept mapping to promote critical thinking. *Journal of Nursing Education, 48,* 45–48.

Wang, H., Sternig, K., Benger, J. & Daya, M. (2015). Randomized trials of prehospital endotracheal intubation in cardiac arrest. *JEMS, 40*(8). Online Mon. Aug 17, 2015 jems.com.

Williams, B. & Brown, T. (2010). Professionalism: Is the Australian paramedic discipline a full profession? *Journal of Emergency Primary Health Care, 8,* 1–10.

Williams, B. & Webb, V. (2015). A national study of paramedic and nursing students' readiness for interprofessional learning (IPL): Results from nine universities. *Nurse Education Today, 35,* 31–37.

Williams, B., Jennings, P.A., Fielder, C. & Ghirardello, A. (2013). Next generation paramedics, agents of change or time for curricula renewal? *Advances in Medical Education and Practice, 4,* 245–250.

Woolard, M. (2012). Professionalism in UK paramedic practice. *Australasian Journal of Paramedicine, 7,* 1–6.

Wyatt, A. (2003). Paramedic practice: Knowledge invested in action. *Australasian Journal of Paramedicine, 1,* 1–9.

7 Teams and Teamwork in Emergency Medical Services

P. Daniel Patterson, Matthew D. Weaver, and David Hostler

CONTENTS

A STORY AND EXAMPLE OF EXCEPTIONAL TEAMWORK IN EMS

When I reflect on exceptional or good teamwork in EMS, I often think of my own experiences working at a small-volume EMS agency that was a mix of paid and volunteer personnel. I typically worked Wednesday evenings for 8 hours. I worked with several different partners. One in particular was my favorite, whom we will call Mitch for confidentiality purposes. He was a great teammate and an experienced paramedic who worked at multiple EMS agencies. Mitch probably worked more than 80 hours a week. Everyone at our agency was friendly, and the medical director was a good friend and mentor. Our agency deployed one ambulance every 24 hours, although we had a second ambulance available if volunteer employees/members were available to respond. My routine with Mitch was to first perform an equipment check, clean the ambulance, and then complete any required documentation or tasks. Next, we would drive the ambulance to a local restaurant near our station, join some colleagues, and enjoy a meal while sharing "war stories" of prior patient encounters. The atmosphere was collegial and fun.

Good teamwork was never a question when I worked with Mitch. We simply worked well together. I recall one shift and one patient encounter as an example of exceptional teamwork. It was a typical Wednesday evening shift. A volunteer

EMT had joined our crew for this evening and sat on the team as a third person. We were familiar with this EMT and enjoyed his presence. On this shift in question, we began our shift by preparing for a continuing education session at a nearby fire department. We were in the ambulance performing our equipment check when the tones (alarm) went off and the county dispatch described an emergency—an older female conscious and alert with respiratory distress. We acknowledged the call and quickly prepared to respond. Mitch and I moved into the cab of our ambulance and instructed the volunteer EMT (whom we will call Iggy) to buckle up in the patient compartment. While I drove, Mitch informed the county dispatch that our unit was en route to the scene with lights and a siren. Mitch and Iggy located the address and provided navigation to the scene.

We arrived on the scene within minutes and found a local police officer at the front door. The officer was a familiar face. He opened the door, which allowed my partners and me to carry an advanced life support house bag, an electrocardiograph (EKG) monitor, oxygen, and a continuous positive airway pressure device into the residence without delay. Once inside, we located our patient. She was in the living room seated upright. She was in obvious respiratory distress, leaning slightly forward and breathing rapidly. She spoke in one- to three-word sentences. I kneeled down and positioned myself directly in front of the patient, looking in her eyes and telling her we were here to help. I reached for her arms and felt her radial pulse. Her skin was warm and pulse rapid, strong, and regular. Mitch and Iggy were positioned to my right (the patient's left side), and proceeded to set up our oxygen supply, the EKG monitor, the intravenous (IV) supplies, and a bag of medications. I interviewed the patient and listened to her lung sounds. The patient described waking suddenly from a nap and experiencing difficulty with breathing. I detected wheezing bilaterally, and before I could request a breathing treatment, my partners had already set up the nebulizer with medication. Before administering the medication, I quickly performed a visual and physical assessment and looked for injury or abnormal physical signs (e.g., edema in lower extremities). I found no physical abnormalities. I confirmed that the patient had no allergies to medication. We immediately began to treat the patient per protocol with breathing treatments and oxygen. Within a few short minutes, we captured a 3- and a 12-lead EKG tracing and detected no evidence of acute cardiac injury. The patient reported "a little bit of chest pain" but denied substernal chest pain or pain that radiated. We decided to follow several treatment protocols simultaneously based on the patient's signs and symptoms. I established an IV in the patient's right antecubital vein and administered additional medication. We also administered aspirin as part of our chest pain protocol. We then moved the patient as a team, belted her securely to the stretcher, and continued with our breathing treatment while loading her into the ambulance for transport. Once in the ambulance, my teammates and I reassessed the patient together. We observed that her respiratory rate had slowed, her work of breathing was improved, and she described relief with additional oxygen. Prior to transport, Mitch and I reviewed our field diagnosis and treatment, considered alternative diagnoses (e.g., a congestive heart failure [CHF] exacerbation and pulmonary embolism), and confirmed pertinent negatives. The patient was stable and showed improvement. We decided that Mitch would drive and Iggy would stay in the patient compartment to help me continue treatment during transport. While traveling to the hospital, I continued to evaluate the patient's EKG tracings, repeated vitals, and provided notification to the receiving facility via a cell phone. We arrived at the hospital within 15 minutes, provided a detailed patient care report, and transferred care with the patient showing signs of improvement.

Later during that same shift, Iggy (the volunteer EMT) had shared with me something he overheard on the scene. He overheard the police officer talking to a neighbor on the scene about our patient care. The officer commented how impressed he was with our quick action and patient treatment. Some days later, our medical director reviewed the patient care report and confirmed that the treatment we delivered was appropriate. In review, this was an ideal prehospital patient care encounter that involved three EMS clinicians working well together—as a team.

When I think of model teamwork in the EMS setting, I often think of the preceding story, as well as other stories that involved other teammates. Unfortunately, not all patient encounters go well, including those with familiar partners. I have many examples where I perceived my actions and my teammate's actions as poor or substandard. This is likely true for many EMS clinicians. We have good and bad experiences with our teammates, and there are many instances where teamwork and team interactions in the EMS setting are simply disorganized. There is a reason to believe that disorganization among teammates contributes to substandard care or negative patient outcomes. Research in non-EMS settings with diverse groups of healthcare and nonhealthcare workers reports that positive teamwork behaviors (e.g., clear and closed-loop communication) and formal training teams can lead to positive outcomes (Salas et al., 2008). Despite the evidence and importance of teamwork and the countless stories of good and bad teamwork, there is limited research devoted to the study of EMS teams. In general, our understanding of EMS teams, their teamwork, and the impact of negative or positive team interactions is poor.

EMS TEAMS

Brannick and Prince (1997) defined a team as two or more people with varying tasks who work interdependently and adaptively to achieve specific and shared goals. Teams of EMS clinicians are in many ways different from teams formed in other organizations. EMS teams are often configured with two clinicians (a dyad) assigned to an ambulance to work in shifts lasting 8, 12, 16, 24 or more hours in duration (Patterson et al., 2011). Their goal is shared and includes care for and stability of the acutely ill or injured and safe transport to a destination appropriate for continuation of patient care. The amount of work that a team performs from shift to shift is unpredictable. One shift may involve one or two critically ill patients, whereas the next shift may require teammates care for five or more patients with non-life-threatening illnesses or injuries. The amount of physical and mental work required varies from patient to patient and from shift to shift. Patients experiencing difficulty breathing as a consequence of CHF often require numerous clinically invasive interventions, delivery of medication, and rapid team coordination. Patients with other conditions may require less action and less team interaction. Teams in other clinical and nonclinical settings are often configured with greater than two teammates, assembling for any number of reasons, ultimately to disband following completion of a specific task or project (Burt & Stevenson, 2009; Harrison, Mohammed, McGrath, Florey & Vanderstoep, 2003; Huckman, Staats & Upton, 2009; Smith-Jentsch, Kraiger, Cannon-Bowers & Salas, 2009; Xu, Carty, Orgill, Lipsitz & Duclos, 2013). An EMS dyad may work on a regular, predictable frequency or may be reconfigured only by chance when individual schedules intersect (Patterson et al., 2011). The size of the EMS organization, shift structure, and other factors influence team/dyadic configuration and duration.

EMS teams often include a combination of any two of the following types of EMS clinicians: first responders, emergency medical technician basics (EMT-Bs), emergency medical technician intermediates (EMT-Is), and emergency medical technician paramedics (EMT-Ps) (Patterson et al., 2011). Less common in ground-based EMS are critical care paramedics (e.g., clinical care EMT-Ps or certified flight paramedics), flight nurses, and EMS physicians (Sucher & Waxler, 2005). Between jurisdictions, there is considerable variation in training, education, scope of practice, and permitted skills for each type of EMS clinician. In the United States, local, regional, state, and national oversight committees/councils govern and establish policies for education and training, scope of practice, and continuing education. EMS clinicians are physician extenders and work under the license of a medical control physician. First responders receive approximately 40 hours of education and training, and their scope of practice is often limited to first-aid-type interventions (e.g., splinting and immobilization, cardiopulmonary resuscitation [CPR]) (Sucher & Waxler, 2005). The EMT-B clinician receives approximately 120 hours of education and training (Sucher & Waxler, 2005). Their scope of practice may go beyond first aid and include trauma care, respiratory, and cardiovascular-related interventions (e.g., insertion of an airway adjunct, bag valve mask ventilation). The EMT-I clinician receives additional education and training beyond that of an EMT-B (Sucher & Waxler, 2005). Their scope of practice may include administration of some medications, obtaining IV access, or other invasive procedures permissible by their protocols and medical control physician. The EMT-P receives 1–2 years of education and training that include clinical rotations in the emergency department (ED), critical care setting, and other clinical locations (Sucher & Waxler, 2005). Paramedics must accumulate hundreds of hours of clinical time and mentorship prior to taking written, oral, and psychomotor tests. A paramedic's scope of practice varies dramatically from EMS system to EMS system, yet often includes invasive procedures such as IVs, endotracheal intubation, medication administration, ventilator management, and other procedures for acutely ill and injured patients.

In addition to the common forms of EMS training, there are various voluntary, specialty certifications offered by independent organizations and advisory committees. A paramedic may obtain additional training and education in the management of critically ill patients in the form of a critical care or flight paramedic certification. Additional training may lead to certification as a tactical paramedic, dive medic, or special operations paramedic. In the United States, physicians often serve in a medical oversight role, developing and approving standing (off-line) protocols and providing consultation in real time (online oversight) when needed. In some locations, physicians may complete a dyad/crew or complement an existing crew. In addition, many EMS systems deploy nurses with prehospital training. EMS teams with nurses are common in the air-medical setting.

Team configuration can differ based on the type of EMS organization and its mission. An EMS system that is integrated within a fire department may configure crews based on standards or policies specific to human resources needs associated with fighting fires and other nonmedical emergencies, allowing the crew to respond to multiple threats within a municipality (Averill et al., 2010; Moore-Merrell et al., 2010). These standards or policies include, but are not limited to, those written by the National Fire Protection Association and the Occupational Safety and Health Administration and recommendations from professional associations or organizations within the industry, such as the International Association of Fire Fighters or the International Association of Fire Chiefs. Fire-based teams may include a crew configuration of two, three, four, or more individuals on a single apparatus (e.g., fire engine or ambulance) (Moore-Merrell et al., 2010). Recent data suggest the configuration of fire-based EMS crews with four fire EMS crew members (to include at least one advanced-level paramedic clinician) as ideal for time-dependent and safe care delivery (Moore-Merrell et al., 2010). Another common delivery is for the fire department to provide a three- or four-person fire suppression crew as first responders, while a separate EMS crew provides the patient transport. This results in a large, multiorganizational EMS team and may present additional challenges to providing quality and efficient patient care.

Team configuration in EMS varies from one EMS system to the next, yet our understanding of this variation is quite limited (Patterson et al., 2011). Much of what is known is based on anecdote and a limited number of studies. Despite the unknown, there are some common features of EMS teams. Many, if not the majority, of EMS systems deploy their crews with at least two EMS clinicians (a dyad). The two clinicians work together on the scene to stabilize the patient and secure them in the patient care compartment of the ambulance. Often only one clinician delivers care en route to the hospital, while the other operates the ambulance. Further complicating the concept is the need for multiple teams to interact while delivering patient care. EMS clinicians may be an element within a larger group such as a search and rescue team. While the group may train together, actual rescue operations are rare and there may be competing demands of the operation and patient care that will challenge the EMS team. Multijurisdictional operations such as a heavy rescue team and an EMS team require two groups who may have no experience working together to communicate effectively to safely perform both rescue/extrication and patient care.

TEAMWORK IN EMS

Teamwork refers to "behaviors of teammates that engender sharing of information and coordination of activities" (Dickinson & McIntyre, 1997). The core components of teamwork include five overarching constructs (team leadership, team orientation, mutual performance monitoring, backup behavior, and adaptability) (Salas, Sims & Burke, 2005). Closed-loop communication, shared mental models (being on the same page), and mutual trust are three mechanisms that help coordinate team constructs (Salas et al., 2005). Conflict among teammates can disrupt or threaten any number of team constructs or coordinating mechanisms and contribute to poor performance or outcomes. There is a limited body of research involving EMS teams, and our understanding of the EMS team/partnership may be described as incomplete (Patterson, Weaver, Weaver et al., 2012; Williams, Rose & Simon, 1999).

Many clinicians will describe teamwork as a vitally important component of EMS work. The nature of EMS work demands positive team-level behaviors performed swiftly with limited information and limited resources. In a typical EMS response, the team will first determine where they need to respond. One crew member will operate the emergency vehicle (frequently at elevated speeds) while his/her teammate gains more information about the dispatch. It is a common expectation that teammates look out for one another. A teammate is expected to warn his/her partner of pending danger, such as approaching traffic. On the scene, teammates are expected to scan the scene and the surrounding area for potential dangers and identify the patient's location. The team is expected to divide responsibilities and carry heavy, awkwardly shaped equipment. At the patient's side, one crew member must act as the team leader and initiate a rapid patient assessment while obtaining information from the patient. The teammate is expected to position needed equipment, obtain baseline vital signs, communicate with bystanders or family members, and call for additional resources as needed.

The preceding description is a simplified characterization of team interaction in the EMS setting. Most patient encounters and the interactions between teammates evolve in a nonlinear fashion. The patient's condition, the scene, and numerous other factors have an impact on the EMS team's actions and teamwork. For many EMS teams, teamwork may occur without verbal direction or communication. With experience, crew members sense or anticipate what is needed, when, where, and how. Nonverbal cues may develop based on a strong knowledge of clinical protocols and teammate familiarity.

Team interactions may become disorganized while treating the acutely ill or injured patient, at scenes where the patient is violent, or in situations with multiple victims or in difficult environments. Teammates are called upon to move quickly and make decisions in haste. This requires that communication be effective and clear. Clinician teammates often experience high levels of stress (Bowron & Todd, 1999; Cydulka, Emerman, Shade & Kubincanek, 1994, 1997; Roth & Moore, 2009), which may lead to poor team interaction and a negative outcome. Teammates routinely lift and move patients over unknown and unfamiliar terrain. Injuries from lifting and moving patients are common among EMS clinicians (Reichard & Jackson, 2010; Reichard, Marsh & Moore, 2011; Suyama, Rittenberger, Patterson & Hostler, 2009). To avoid risk of injury, a team's actions for the purpose of lifting and moving patients or equipment require coordination and communication. Miscommunication may contribute to patient or clinician injury (e.g., a muscle strain or sprain or a patient drop from a stretcher) (Suyama et al., 2009; Wang, Weaver, Abo, Kaliappan & Fairbanks, 2009). Team communication is particularly important while moving a patient from the scene to the ambulance. This process alone requires continuous coordination of movements by the team, steering and moving the stretcher across uneven terrain, to eventually be secured into the back of the ambulance. Once inside, the team must again coordinate their actions to reassess the patient for any change, reattach any equipment or monitoring devices applied on the scene, draw up and administer medications, and perform other life-saving interventions. One crew member must remain with the patient to care for the patient while en route to the hospital while his/ her teammate operates the ambulance. If emergent transport is required, the ambulance operator may proceed through various traffic controls without the assistance of his/her teammate monitoring the route for threats to safety. The ambulance operator must continue to look out for his/her teammate in the patient compartment and warn his/her partner (*with a loud voice*) of pending danger to him/her and the patient (e.g., bumps in the road, abrupt stops and starts due to traffic). Teamwork continues with patient movement into the hospital, where care is transferred to the receiving clinicians following a verbal report and physical movement of the patient from the EMS stretcher onto the hospital stretcher. The team continues to work together to complete a patient care report, clean and restock the ambulance, and safely return to service.

PERFORMANCE METRICS FOR EMS TEAMS

Standard measures of EMS team performance are lacking, which makes it difficult to quantitatively assess EMS team performance. Few studies have examined the performance of EMS teams with the goal of linking teamwork characteristics to outcomes. The performance of EMS systems, not teams, has been a leading measure of interest (Moore, Swor & Pirrallo, 2005; Myers et al., 2008; Rahman et al., 2015).

Previously investigated indicators of team or individual performance include time or speed to intervention or completion of steps in sequence of a protocol (Bayley, Weinger, Meador & Slovis, 2008; Eschmann, Pirrallo, Aufderheide & Lerner, 2010; Martin-Gill, Guyette & Rittenberger, 2010). Some have investigated these indicators in simulated settings with patient care scenarios of cardiac arrest (Bayley et al., 2008; Eschmann et al., 2010; Martin-Gill et al., 2010). Bayley et al. (2008) investigated the impact of crew configuration on performance in simulated cases of cardiac arrest. Performance measures included errors of commission, errors of omission, errors of sequence, and compliance with chest compressions (or hands-on/hands-off time). Errors of omission refer to failure to perform a step or key event of a protocol (e.g., administration of a medication). Errors of commission refer to the act of doing something out of sequence or performing an intervention outside of protocol that is unnecessary or potentially harmful. Martin-Gill et al. (2010) investigated the impact of diverse crew configurations on performance in simulated cases of cardiac arrest. Investigators used no-flow fraction (NFF) as their primary outcome measure of interest; NFF is a surrogate for low coronary perfusion pressure and an indicator of quality CPR. Eschmann et al. (2010) examined the association between EMS team factors and performance in a retrospective study with more than 10,000 patient care charts involving cardiac arrest. Their team factor of interest was the number of paramedics on scene (2 versus 3 versus 4+ paramedics), and their outcomes of interest were return of spontaneous circulation and survival to hospital discharge (Eschmann et al., 2010).

Evaluation of team interaction and performance in the EMS setting is a challenge. EMS teams change their location frequently and randomly based on need and demand. This poses a problem for observation of team behavior and interaction in the natural environment. Observation is key to the measurement of teams and their interaction, behavior, and performance (Baker & Gallo, 2013). Evaluation of EMS teams is also challenged by the lack of an established definition of EMS teams and teamwork and limited testing of team measurement tools (Patterson, Weaver, Weaver et al., 2012; Williams et al., 1999). The need for reliable, valid, and widely tested measures of EMS team performance is compelling.

THREATS TO EMS PERFORMANCE

Despite the lack of team performance metrics, EMS teams face numerous threats that may impact individual and team-level performance including familiarity, fatigue, and the organization's safety culture.

TEAMMATE FAMILIARITY

Team familiarity refers to the amount of time two or more individuals spend together (are exposed to one another) over time (Patterson, Pfeiffer et al., 2015). Previous research shows a relationship between familiarity, behavior, and outcomes. One study by the National Aeronautics Space Administration detected a greater occurrence of errors and communication unrelated to work among unfamiliar teammates versus familiar teammates (Foushee, Lauber, Baetge & Acomb, 1986). A study of pilot/copilot teams showed a higher rate of errors during the takeoff and landing phases of a flight than among familiar crew members (Thomas & Petrilli, 2006). Similar associations in familiarity and measures of performance have been detected in other occupations (Harrison et al., 2003; Huckman et al., 2009; Xu et al., 2013).

Few studies have examined teammate familiarity in EMS (Patterson et al., 2011; Patterson, Pfeiffer et al., 2015; Patterson, Vena et al., 2015). The available data show that few EMS clinicians spend much of their time at work with a regular teammate. One study of three diverse EMS organizations determined that EMS clinicians work with 19 different partners annually and spend approximately 35% of their time at work with the same partner (Patterson et al., 2011). A more recent study of more than 60,000 unique EMS dyads showed that limited familiarity with a teammate could lead to injury (Patterson, Vena et al., 2015). In this study, the incidence of injury was highest for dyads with the lowest amount of familiarity (Patterson, Vena et al., 2015). Compared to dyads on their first shift together, dyads that had worked together two to three times were nearly 60% less likely to report an injury (incidence rate ratio [IRR] 0.43, 95% confidence interval [CI] 0.31–0.60). Dyads who worked at least 10 shifts together over the study period had a far lower risk of injury (IRR 0.03 (95% CI 0.02–0.04).

FATIGUE

Fatigue refers to "a subjective, unpleasant symptom, which incorporates total body feelings ranging from tiredness to exhaustion creating an unrelenting overall condition which interferes with an individual's ability to function to their normal capacity" (Ream & Richardson, 1996). More than half of EMS clinicians report severe mental and/or physical fatigue while at work (Patterson et al., 2014; Patterson, Suffoletto, Kupas, Weaver & Hostler, 2010; Patterson, Weaver, Frank et al., 2012). Odds of injury and medical error are greater among fatigued EMS clinicians than nonfatigued clinicians (Patterson, Weaver, Frank et al., 2012). Excessive fatigue limits physical strength, reduces a person's reaction time, and negatively impacts decision-making (Barker & Nussbaum, 2011; Killgore, Grugle & Balkin, 2012; Lim & Dinges, 2008; Patterson, Weaver, Frank et al., 2012). Individuals who are fatigued have difficulty processing information and adapting to changing circumstances (Whitney, Hinson, Jackson & Van Dongen, 2015)—making positive teamwork crucial in the EMS environment. Unfortunately, teammates may behave differently when fatigued and fail to engage in positive team behaviors. Fatigued clinicians may not communicate as well or fail to provide a teammate backup assistance when needed. The effectiveness of fatigue countermeasures in relation to teamwork is unclear. The interaction between teamwork and fatigue-related impairment is also poorly understood and deserves further study.

SAFETY CULTURE

Safety culture refers to "the shared perceptions or attitudes of a work group toward safety" (Zohar, 1980). Measurement of safety culture involves assessment of six distinct domains, including "teamwork climate" (Patterson, Huang, Fairbanks & Wang, 2010). In prior research, we showed that teamwork climate varies widely among EMS agencies (Patterson, Huang, Fairbanks, Simeone et al., 2010). Among 61 EMS agencies distributed across the United States, scores for teamwork climate ranged from 45 to 90 on a 100-point scale (Patterson, Huang, Fairbanks, Simeone et al., 2010). The mean score for teamwork climate was 71. Scores >75 indicate positive perceptions of teamwork. In a follow-up study of 21 EMS organizations, we examined the linkage between teamwork climate and safety outcomes (Weaver, Wang, Fairbanks & Patterson, 2012). The mean score for teamwork climate (70) remained below the cut point used to indicate "positive perceptions of teamwork." We also determined that lower scores on teamwork climate were independently associated with higher odds of self-reported injury, medical errors or adverse events, and safety compromising behaviors (Weaver, Wang et al., 2012). These data show that many EMS clinicians perceive teamwork in EMS organizations as less than positive and poor perceptions are associated with negative outcomes. Despite these data, we have a limited understanding of teamwork climate and the role it plays in team behavior, team performance, and safety.

IMPROVING TEAMWORK IN EMS AND CONCLUDING REMARKS

Team performance may be enhanced by training teammates to know the roles and responsibilities of his/her teammates and training individuals to perform tasks of teammates. Team performance is improved by promoting development of teammates of shared mental models and use of these models to coordinate actions, adapt to changing environments, and predict when teammates need help. Guiding team-mates through a process of identifying threats to teamwork and taking corrective action in real time and through a retrospective review process is impactful (Driskell, Lazzara, Salas, King & Battles, 2013).

Despite supportive evidence for team training, all teams are not created equal, and team training has not been evaluated in all types of teams (Weaver, Feitosa, Salas, Seddon & Vozenilek, 2013). Teams in the EMS environment are unique and face challenges not often seen in other settings, including other healthcare settings such as the ED. EMS teams work in small groups (most often a dyad) and under con-ditions that change frequently and in environments that are often unfamiliar. Many EMS teammates are not accustomed to working with a steady partner (Patterson et al., 2011), many are fatigued and under a great deal of stress, and many work in a poor organizational culture. These factors as well as other elements pose a chal-lenge for building and sustaining positive teamwork in the EMS environment. We cannot begin to fix problems with EMS team performance until the problem(s) have been adequately defined and described. More research (observational research to be specific) is needed. However, novel approaches to training EMS teams are also needed and the lack of research should not prevent development of novel approaches to teamwork assessment and intervention.

Team training should begin in the EMS education and initial training environ-ment. Currently, there is no established mechanism or program that educates EMS students on the common/core principles of teamwork. In fact, students are evaluated almost exclusively on individual competencies. Terminal performance objectives, or the number of successful skills (e.g., IV access) required to complete training, are largely derived by consensus. In certain programs, an EMS student may be required to "lead the team" consisting of the preceptor and his or her partner but there is rarely an evaluation of how the student integrates into the team. Even the skill test-ing required for initial certification is evaluated for each individual, and there is no formal evaluation of teamwork.

Where there has been a focus on team performance, this focus is almost exclu-sively on low-frequency, high-acuity patient encounters (e.g., cardiac arrests). This is despite the fact that greater than 15% of EMS encounters involve patients with conditions that may be otherwise classified as stable, nonurgent, nonemergent, or medically unnecessary (Alpert, Morganti, Margolis, Wasserman & Kellerman, 2013; Weaver, Moore, Patterson & Yealy, 2012). This does not mean that 85% are high-acuity events. In fact, the proportion of high-acuity events is relatively small. Successful execution of low-frequency/high-consequence events such as cardiac arrest requires considerable coordination of teammates, rapid delivery of appropri-ate therapies, and vigilance by all individuals involved. Patient outcomes have been linked to CPR that is performed without interruption and in sequence with cur-rent protocol. As a result, local, state, and national EMS oversight organizations have modified protocols and emphasized greater coordination, communication, and quick action among EMS teams involved in out-of-hospital cardiac arrest resuscita-tion. There is limited evidence of similar efforts to enhance team performance other than for out-of-hospital cardiac arrest and no studies examining a potential spillover effect from team training for cardiac arrest into other situations.

Several strategies for teamwork improvement that may be implemented with limited impact on established work processes include the following:

1. Require newly formed EMS crews to arrive 20–30 minutes prior to the start of a shift. Have crew members rehearse who will do what on the scene. Require crew members to discuss strengths and weaknesses and to offer a plan for how one teammate will work to resolve a weakness identified by the other teammate.
2. Use simulated patient cases during scheduled continuing education sessions when communication and other teamwork behaviors are stressed. Have teammates identify deficits in teamwork and offer strategies that enhance teamwork. No single strategy may work well for every team formed. Have crew members commit to practicing these strategies when partnered for shift work.
3. Introduce standardized team training programs as a regular component of new employee training (onboarding) and continuing education.

Team training works (Salas et al., 2008). Unfortunately, little attention has been devoted to the "team element" of EMS operations and care delivery. Team performance, quality of care, and safety in EMS can be improved with team training that is woven into the fabric of initial and continuing education.

REFERENCES

Alpert, A., Morganti, K. G., Margolis, G. S., Wasserman, J. & Kellerman, A. L. (2013). Giving EMS flexibility in transporting low-acuity patients could generate substantial Medicare savings. *Health Affairs, 32*(12), 2142–2148.

Averill, J. D., Moore-Merrell, L., Barowy, A., Santos, R., Peacock, R., Notarianni, K. A. et al. (2010). *Report on residential fireground field experiments.* Retrieved from: http://www.nist.gov/el/fire_research/upload/Report-on-Residential-Fireground-Field-Experiments.pdf.

Baker, D. P. & Gallo, J. (2013). Measuring and diagnosing team performance. In D. P. Baker, J. B. Battles, H. B. King & R. L. Wears (Eds.), *Improving patient safety through teamwork and team training* (pp. 234–238). New York: Oxford University Press.

Barker, L. M. & Nussbaum, M. A. (2011). Fatigue, performance and the work environment: A survey of registered nurses. *Journal of Advanced Nursing, 67*(6), 1370–1382.

Bayley, R., Weinger, M., Meador, S. & Slovis, C. (2008). Impact of ambulance crew configuration on simulated cardiac arrest resuscitation. *Prehospital Emergency Care, 12*(1), 62–68.

Bowron, J. S. & Todd, K. H. (1999). Job stressors and job satisfaction in a major metropolitan public EMS service. *Prehospital Disaster Medicine, 14*(4), 236–239.

Brannick, M. T. & Prince, C. (1997). An overview of team performance measurement. In M. T. Brannick, E. Salas & C. Prince (Eds.), *Team performance assessment and measurement* (pp. 3–16). Mahwah, NJ: Lawrence Erlbaum Associates.

Burt, C. D. & Stevenson, R. J. (2009). The relationship between recruitment processes, familiarity, trust, perceived risk and safety. *Journal of Safety Research, 40*(5), 365–369.

Cydulka, R. K., Emerman, C., Shade, B. & Kubincanek, J. (1994). Stress levels in EMS personnel: A longitudinal study with work-schedule modification. *Academic Emergency Medicine, 1*(3), 240–246.

Cydulka, R. K., Emerman, C. L., Shade, B. & Kubincanek, J. (1997). Stress levels in EMS personnel: A national survey. *Prehospital and Disaster Medicine, 12*(2), 136–140.

Dickinson, T. L. & McIntyre, R. M. (1997). A conceptual framework for teamwork measurement. In M. T. Brannick, E. Salas & C. Prince (Eds.), *Team performance assessment and measurement* (pp. 19–43). Mahwah, NJ: Lawrence Erlbaum Associates.

Driskell, T., Lazzara, E. H., Salas, E., King, H. B. & Battles, J. B. (2013). Does team training work? Where is the evidence? In E. Salas, K. Frush, D. P. Baker, J. B. Battles, H. B. King & R. L. Wears (Eds.), *Improving patient safety through teamwork and team training* (pp. 201–217). New York: Oxford University Press.

Eschmann, N. M., Pirrallo, R. G., Aufderheide, T. P. & Lerner, E. B. (2010). The association between emergency medical services staffing patterns and out of hospital cardiac arrest survival. *Prehospital Emergency Care, 14*(1), 71–77.

Foushee, H. C., Lauber, J. K., Baetge, M. M. & Acomb, D. B. (1986). *Crew factors in flight operations III: The operational significance of exposure to short-haul air transport operations* (Report No. 88322). Moffett Field, CA: National Aeronautical and Space Administration.

Harrison, D. A., Mohammed, S., McGrath, J. E., Florey, A. T. & Vanderstoep, S. W. (2003). Time matters in team performance: Effects of member familiarity, entrainment, and task discontinuity on speed and quality. *Personnel Psychology, 56*(3), 633–669.

Huckman, R. S., Staats, B. R. & Upton, D. M. (2009). Team familiarity, role experience, and performance: Evidence from Indian software services. *Management Science, 55*(1), 85–100.

Killgore, W. D., Grugle, N. L. & Balkin, T. J. (2012). Gambling when sleep deprived: Don't bet on stimulants. *Chronobiology International, 29*(1), 43–54.

Lim, J. & Dinges, D. F. (2008). Sleep deprivation and vigilant attention. *Annals of the New York Academy Science, 1129*(1), 305–322.

Martin-Gill, C., Guyette, F. & Rittenberger, J. C. (2010). Effect of crew size on objective measures of resuscitation for out-of-hospital cardiac arrest. *Prehospital Emergency Care, 14*(2), 229–234.

Moore, L., Swor, R. A. & Pirrallo, R. G. (2005). Performance measurement in EMS. *Improving Quality in EMS*. Dubuque, IA: Kendall/Hunt.

Moore-Merrell, L., Santos, R., Wissoker, D., Benedict, R., Taylor, N., Goldstein, R. et al. (2010). *Report on EMS field experiments: Firefighter safety and deployment study*. Retrieved from: https://www.iaff.org/tech/PDF/EMS%20Nist%20Report_LORES.pdf.

Myers, J. B., Slovis, C. M., Eckstein, M., Goodloe, J. M., Isaacs, S. M., Loflin, J. R. et al. (2008). Evidence-based performance measures for emergency medical services systems: A model for expanded EMS benchmarking. *Prehospital Emergency Care, 12*(2), 141–151.

Patterson, P. D., Arnold, R. M., Abebe, K., Lave, J. R., Krackhardt, D., Carr, M. et al. (2011). Variation in emergency medical technician partner familiarity. *Health Services Research, 46*(4), 1319–1331.

Patterson, P. D., Buysse, D. J., Weaver, M. D., Suffoletto, B. P., McManigle, K. L., Callaway, C. W. & Yealy, D. M. (2014). Emergency healthcare worker sleep, fatigue, and alertness behavior survey (SFAB): Development and content validation of a survey tool. *Accident Analysis Prevention, 73*(1), 399–411.

Patterson, P. D., Huang, D. T., Fairbanks, R. J., Simeone, S. J., Weaver, M. D. & Wang, H. E. (2010). Variation in emergency medical services workplace safety culture. *Prehospital Emergency Care, 14*(4), 448–460.

Patterson, P. D., Huang, D. T., Fairbanks, R. J. & Wang, H. E. (2010). The emergency medical services safety attitudes questionnaire. *American Journal of Medical Quality, 25*(2), 109–115.

Patterson, P. D., Pfeiffer, A. J., Lave, J. R., Weaver, M. D., Abebe, K., Krackhardt, D. et al. (2015). How familiar are clinician teammates in the emergency department? *Emergency Medicine Journal, 32*(4), 258–262.

Patterson, P. D., Suffoletto, B. P., Kupas, D. F., Weaver, M. D. & Hostler, D. (2010). Sleep quality and fatigue among prehospital providers. *Prehospital Emergency Care, 14*(2), 187–193.

Patterson, P. D., Vena, J., Weaver, M. D., Hostler, D., Salas, E., Krackhardt, D. et al. (2015). Association between EMS teammate familiarity and injury. *Prehospital Emergency Care, 19*(1), 151.

Patterson, P. D., Weaver, M. D., Frank, R. C., Warner, C. W., Martin-Gill, C., Guyette, F. X. et al. (2012). Association between poor sleep, fatigue, and safety outcomes in emergency medical services providers. *Prehospital Emergency Care, 16*(1), 86–97.

Patterson, P. D., Weaver, M. D., Weaver, S. J., Rosen, M. A., Todorova, G., Weingart, L. R. et al. (2012). Measuring teamwork and conflict among emergency medical technician personnel. *Prehospital Emergency Care, 16*(1), 98–108.

Rahman, N. H., Tanaka, H., Shin, S. D., Ng, Y. Y., Piyasuwankul, T., Lin, C. H. et al. (2015). Emergency medical services key performance measurement in Asian cities. *International Journal of Emergency Medicine, 8*(12).

Ream, E. & Richardson, A. (1996). Fatigue: A concept analysis. *International Journal of Nursing Studies, 33*(5), 519–529.

Reichard, A. A. & Jackson, L. L. (2010). Occupational injuries among emergency responders. *American Journal of Industrial Medicine, 53*(1), 1–11.

Reichard, A. A., Marsh, S. M. & Moore, P. H. (2011). Fatal and nonfatal injuries among emergency medical technicians and paramedics. *Prehospital Emergency Care, 15*(4), 511–517.

Roth, S. G. & Moore, C. D. (2009). Work–family fit: The impact of emergency medical services work on the family system. *Prehospital Emergency Care, 13*(4), 462–468.

Salas, E., DiazGranados, D., Klein, C., Burke, C. S., Stagl, K. C., Goodwin, G. F. et al. (2008). Does team training improve team performance? A meta analysis. *Human Factors, 50*(6), 903–933.

Salas, E., Sims, D. E. & Burke, C. S. (2005). Is there a "Big Five" in teamwork? *Small Group Research, 36*(5), 555–599.

Smith-Jentsch, K. A., Kraiger, K., Cannon-Bowers, J. A. & Salas, E. (2009). Do familiar teammates request and accept more backup? Transactive memory in air traffic control. *Human Factors, 51*(2), 181–192.

Sucher, M. A. & Waxler, J. L. (2005). EMS providers and system roles. In J. Brennan & J. Krohmer (Eds.), *Principles of EMS systems.* Sudbury, MA: Jones & Bartlett.

Suyama, J., Rittenberger, J. C., Patterson, P. D. & Hostler, D. (2009). Comparison of public safety provider injury rates. *Prehospital Emergency Care, 13*(4), 451–455.

Thomas, M. J. & Petrilli, R. M. (2006). Crew familiarity: Operational experience, non-technical performance, and error management. *Aviation Space Environmental Medicine, 77*(1), 41–45.

Wang, H. E., Weaver, M. D., Abo, B. N., Kaliappan, R. & Fairbanks, R. J. (2009). Ambulance stretcher adverse events. *Quality and Safety in Health Care, 18*(3), 213–216.

Weaver, S. J., Feitosa, J., Salas, E., Seddon, R. & Vozenilek, J. A. (2013). The theoretical drivers and models of team performance and effectiveness for patient safety. In E. Salas, K. Frush, D. P. Baker, J. B. Battles, H. B. King & R. L. Wears (Eds.), *Improving patient safety through teamwork and team training* (pp. 3–26). New York: Oxford University Press.

Weaver, M. D., Moore, C. G., Patterson, P. D. & Yealy, D. M. (2012). Medical necessity in emergency medical services transports. *American Journal of Medical Quality, 27*(3), 250–255.

Weaver, M. D., Wang, H. E., Fairbanks, R. J. & Patterson, P. D. (2012). The association between EMS workplace safety culture and safety outcomes. *Prehospital Emergency Care*, *16*(1), 43–52.

Whitney, P., Hinson, J. M., Jackson, M. L. & Van Dongen, H. P. (2015). Feedback blunting: Total sleep deprivation impairs decision making that requires updating based on feedback. *Sleep*, *38*(5), 745–754.

Williams, K. A., Rose, W. D. & Simon, R. (1999). Teamwork in emergency medical services. *Air Medical Journal*, *18*(4), 149–153.

Xu, R., Carty, M. J., Orgill, D. P., Lipsitz, S. R. & Duclos, A. (2013). The teaming curve: A longitudinal study of the influence of surgical team familiarity on operative time. *Annals of Surgery*, *258*(6), 953–957.

Zohar, D. (1980). Safety climate in industrial organizations: Theoretical and applied implications. *Journal of Applied Psychology*, *65*(1), 7.

8 Defining the Prehospital Care Multiteam System*

Deborah DiazGranados,
Marissa L. Shuffler, Nastassia Savage,
Alan W. Dow, and Harinder Dhindsa

CONTENTS

A 21-year-old male driver was driving home while intoxicated. Rounding a curve, he skidded off the road into a fence and flipped his car, which then caught fire due to a gasoline leak. The 911 operator dispatched the police, the fire department, and an ambulance. Each of these units, rapidly responding to the scene, had specific responsibilities—the dispatcher to provide prearrival instructions to bystanders; the police to secure the scene for public safety and criminal investigation; and the fire department to extinguish the fire, extricate the patient from the car, and protect other public safety responders. The responsibilities of the ambulance crew, or emergency medical service, included performance of a rapid patient assessment, initiation of resuscitative and stabilization measures, expedient transport to and prearrival communication with the nearest trauma center. What principles guide the work of the emergency medical service as they try to complete their work? How does the core team integrate and coordinate with other individuals or teams to provide effective service to meet the needs of society and individuals?

---

* The content is solely the responsibility of the authors and does not necessarily represent the official views of NCATS, NIH, or GHS.

INTRODUCTION

The example provided above is not an anomaly in terms of the scenarios that EMS personnel experience. The cases are complex, with a number of professionals involved in emergency response. Take, for example, the initial response to the car crash in this case. Emergency response typically involves police, fire, and EMS individuals, each of whom is guided by a specific mission. The police serve the societal need for order and justice. They are responsible for enforcing laws through investigation and referral of suspected lawbreakers to the legal system. Helping individuals is an important part of the police role, but their broader responsibility and professional charge come from society as a whole. Fire professionals serve a different societal need by protecting citizens from natural harm. While fire-related emergencies are the classic example, fire professionals execute that mission in a myriad of ways ranging from limiting the dangers to other professionals at a car crash, performing complicated rescues, and operating with hazardous materials to rescuing the proverbial kitten from a tree. In comparison to the police, fire professionals serve needs more driven by the individual, albeit in a manner defined and financed by society as a whole. EMS professionals also serve a societal need to provide emergency care, yet this is perhaps the most personal of all three professional groups. By its very nature, EMS workers are commonly dealing with life-or-death situations. Their goal is not maintaining societal order or preventing natural harm but rather preserving life, regardless of the attendant circumstance. As such, police, fire, and EMS professionals work in a world of complexity where different missions, professions, and structures for work charges must meld to provide citizens the expected services.

The environment wherein EMS professionals function can be described as hypercomplex. An environment in which hypercomplexity exists is one with an extreme variety of components, systems, and levels, each having distinct training, leadership, and operational standards (Roberts & Rousseau, 1989). When systems, components, or levels in healthcare must be interdependent, it is essential that they coordinate their activities and, in the case of EMS, effectively provide emergency care to patients. The varied component teams (e.g., 911, fire department, ambulance, hospital) must, at the basic level, operate interdependently by communicating information to each other based on the patient's condition and their physical assessments. Moreover, for these component teams to be effective, they must be tightly coupled. This hypercomplex world requires teamwork within and between the teams that comprise the multiteam system (MTS).

MTSs are a complex hybrid organizational form. The complexity, which is characteristic of an MTS, requires individual teams who are part of the large system to plan and set goals while still being flexible enough to adjust, if needed, to the goals and plans for the MTS as a whole (DeChurch & Mathieu, 2009; Davison et al., 2012). Part of the troubles that EMS face is that coordination between component teams is often inadequate. In a study of EMS, it was reported that time pressures, competing demands, and a lack of trust are often hindrances to this coordination (Institute of Medicine, 2006). Moreover, the Institute of Medicine (2006) report cites that challenges to prehospital care and to coordination are exacerbated by ED crowding, the challenge of training a workforce with the healthcare expertise and interpersonal skills in coordination, and maintaining consistency within an EMS system. Very little scientific evidence exists to inform our understanding of prehospital care and EMS, particularly in the context of MTSs.

In the example that begins this chapter, personnel from police, fire, and emergency services represent an MTS. Each group contains highly trained experts with an understanding of who is responsible for which tasks. In addition, authority for the overall scene is determined by the degree of need for a specific group's expertise. For example, at a crash scene involving an uncontrolled fire, unsafe for rescue personnel as well as crash victims, fire professionals would lead the initial efforts. Building from this example, we draw from the definition of an MTS as well as new research within the organizational science literature to examine the challenges faced by the broad-based and collaborative activities in which prehospital care EMS professionals engage and provide some direction for further study and enhancement of the existing system. We offer a discussion of defining characteristics of the EMS MTS and discuss the challenges that may result from these defining characteristics. Moreover, in this chapter, we offer an example, in more detail than the broad example provided at the beginning of this chapter, of EMS as an MTS so that the challenges can be highlighted and discussed with a level of specificity. We posit that by understanding how EMS professionals work within the organizational context of an MTS, improvements can be made to operational procedures and training within and between varied components, systems, and levels of the MTS to reduce and mitigate the challenges to coordination, team identity, leadership, and communication.

WHAT IS AN MTS?

Through observations in real-world organizational settings, Mathieu et al. (2001) were able to discern a unique and specific grouping of employees that was not explained by team and organizational levels of analysis, which they termed an MTS. An MTS is defined as "two or more teams that interface directly and interdependently in response to environmental contingencies toward the accomplishment of collective goals" (Mathieu et al., 2001, p. 290). Put another way, an MTS is a network of teams that work together toward at least one collective goal while simultaneously working toward the completion of individual team goals (Shuffler et al., 2014). These types of systems have been found in emergency disaster responses such as those to Hurricanes Andrew and Katrina (Moynihan, 2007; DeChurch et al., 2011), in the military such as with provincial reconstruction teams (Goodwin et al., 2012), and, as is the focus of this book, in healthcare with the treatment of mass casualty victims and emergency room patients, such as a car crash victim as described in the opening example. As we have illustrated, MTSs exist in a variety of contexts in which multiple groups must work together to achieve a common goal.

Assessing individuals in terms of MTSs allows for an explanation of the nuances that occur in these systems, for which traditional research on organizations and teams does not account (Mathieu, 2012). There is a unique set of information that can be gathered by assessing these systems of teams, which require a high level of coordination among the distinct but interdependent component teams while working together toward a critical goal, such as saving the lives of patients (DeChurch et al., 2011). Yet this shared goal may not be the only goal of each component team. For example, while EMS serves the public good through emergent health services, the police serves the public good through enforcing law. In the opening example, enforcing law by arresting the drunk driver is in conflict with the need for evaluation of his injuries. Here, the solution for coordination between these groups is clear—triage and treat injuries on the scene or at an emergency department (ED) then arrest—but what if the injuries are extremely minor? Or what if the criminal act involves a death? Depending on the circumstances, the solution to coordination problems can become less clear.

As MTSs can and do occur in various organizational, military, and medical contexts, it is important to consider how to distinguish between different types of MTSs. Zaccaro et al. (2012) classified the different characteristics into three sets of attributes: compositional, linkage, and developmental (see Table 8.1). Compositional attributes refer to the demographic and descriptive characteristics of the MTS and its component teams. These characteristics include the size of the MTS, the number of component teams, the extent to which they come from different organizations, the differences in their core goals, the degree to which they are geographically dispersed, how diverse the component teams are, the extent that their goals align with those of the MTS, and the amount of time and effort dedicated to the MTS. Linkage attributes focus on the different types of linking mechanisms that connect component teams to one another in an MTS. These characteristics include the degree of coordination between teams, their level of responsibility, the power distribution, and the structure communication. The third attribute is developmental and refers to the characteristics that depict how MTSs are shaped and formed over time. These characteristics include whether the MTS was appointed or emerged naturally, the anticipated life span of the MTS, the stage of formation in the MTS (e.g., newly formed or mature), and the fluidity or stability in membership and communication.

Many MTSs operate through strong task-related linkages among component teams such that the component teams understand when to communicate with one another (Mathieu et al., 2001). For example, car crashes always involve response

TABLE 8.1
MTS Typology

Attribute	Definition	Examples
Compositional	Demographic and descriptive characteristics of the MTS and its component teams	• Number of component teams in MTS • Size of the MTS • Diversity in the MTS
Linkage	Describe the different types of linking mechanisms that connect the component teams	• Interdependence among teams • Level of responsibility among teams • Distribution of power within the MTS
Developmental	Depict how the MTS was shaped or formed over time	• Appointment or emergence of MTS • Expected duration of MTS • Stability of membership and communication in MTS

from fire, police, and EMS professionals and leadership at the scene depends on the specific factors of the environment. MTSs also tend to work outside of organizational boundaries (Shuffler et al., 2014), such that an MTS tends to incorporate teams from various departments and backgrounds to complete their shared goal(s). One example of this is the various medical teams that assist in the treatment of a cancer patient, including a medical oncology team, a surgical team, a pathology team, and a radiology team. In a high-functioning cancer center, these disciplines collaborate to deliver the most appropriate care for each patient. Because of the context in which MTSs occur, they are often considered a task situation such that their existence is reliant upon the completion of this task (i.e., cancer remission) and effort among and across the component teams is organized and managed above the team or individual level, which can occasionally occur across multiple organizations (Mathieu et al., 2001). In the following section, this understanding of an MTS is applied to the context of EMS.

MTSs IN EMS RESPONSES

EMSs aim to provide effective emergency medical care to those who require it. The core team always includes first responders trained in the delivery of healthcare services such as paramedics and EMTs. In some countries, nurses, physicians, and other health professions may also be part of the first responder team. In addition, the team may also include dispatchers who initiate and help coordinate the work of EMS as well as the doctors, nurses, and specialists who may also help coordinate care and provide care at hospitals or other medical facilities (Kobusingye et al., 2006). Within this chain between dispatchers and medical facilities, EMS must respond quickly to evaluate the patient's situation, rescue and extricate the patient if indicated, and provide immediate care.

The composition of the team and the structure of the local healthcare system define the linkages and work processes of the EMS team. These work processes may occur on the scene by the EMTs or paramedics and may not require any further treatment; however, additional care may be needed and patients may be transported to the nearest hospital. At this point, the EMS personnel would communicate the incoming patient to the hospital and the ED would prepare for their arrival based upon the patient's status as communicated by EMS. In addition, the added expertise of physicians and other health professionals may guide additional care in the field through both off-line and online medical direction. Off-line medical direction refers to the patient care protocols that EMS personnel use and that typically fall under the responsibility of the physician medical director for the EMS agency. Online medical direction, on the other hand, refers to guidance and direction that EMS personnel receive in real time, usually from an ED physician at the hospital or, at times, from the agency medical director.

As an example, in the case of a patient with chest pain, an EKG may be obtained and aspirin administered per agency protocol (off-line medical direction), and in the case of a patient with a heart attack, they may receive online direction to bypass the local hospital and go to the nearest hospital with a cardiac catheterization lab that is open in order to provide more specialized care to the patient (online medical direction). Once the patient arrives, the personnel of the ED then assumes care for the patient and provides additional assessment, stabilization, and treatment as needed. The patient may be released from the hospital or transferred to other teams or inpatient services for further observation and treatment (DiazGranados et al., 2014). EMSs work as part of an MTS across the different departments with both concurrent (initial response with fire and police) and sequential (transfer of care from 911 to EMS to hospital) needs for coordination to ensure that the patient receives appropriate care.

The compositional and linkage attributes of MTSs in healthcare depend on the acuity or temporality of each patient's case. For instance, a cancer patient has a significant medical condition with ongoing needs for care. This patient's MTS likely includes oncologists, radiologists, pharmacists, and many other teams working to treat the cancer and maintain the patient's quality of life during treatment. In this scenario, the compositional characteristics of the teams (size and number) might be relatively large. Because cancer is unfortunately common, the linkage of these teams should be strong and is often formalized through meetings or referral protocols. Finally, the consistency of team members and teams supports a structured developmental process that seeks to improve the process of care and patient outcomes.

In contrast, consider a patient who arrives at the emergency room with a broken arm. This patient would require fewer departments—radiology and orthopedics—and less ongoing time and effort than the cancer patient. While the linkage between the ED and the other services may have defined referral patterns, the specific healthcare providers may not have collaborated in the care of patients before. Organizational policies might govern coordination but might also hinder coordination by not being appropriate for the specifics of this type of collaboration. For example, a policy requiring urgent notification by the radiologist of significant radiographic findings such as an acute fracture may not be relevant when the orthopedist has already viewed the images in the electronic system as well. Developmentally, this MTS forms and dissolves over hours. The care of the patient—whether discharged home or admitted—is assumed by the orthopedist, while the ED physician and radiologist have no ongoing role. As such, activities that might improve future performance of the team—training and formalization of procedures—may not be of benefit, even though this particular MTS might be needed again in the future.

A smaller MTS has less opportunity for diversity to impact its work, be it across organizations, cultures, or functions and backgrounds. One of the compositional challenges in an MTS is balancing the benefits of diverse perspectives and expertise with the detriments of additional complexity and the inconsistencies across various component team processes. Diversity can have advantages by adding new ideas and expertise and disadvantages through increased needs for coordination and effort toward team formation. EMSs require functional diversity such that they can treat the ailment(s) of the patients that they receive. Yet when component teams are geographically dispersed from one another, additional efforts must be made to ensure coordination on account of the distribution and time differences. Differences in the level of commitment and effort that interfacing parts of the system can dedicate to each patient may vary depending upon the severity of the case and the workload within each part of the system. For example, if an ED is overcrowded and/or understaffed, they may be more inclined to more rapidly transfer a patient to inpatient services to free a bed in their own department (DiazGranados et al., 2014). The impact of diversity—at the core team and system level—on performance is an area ripe for further research.

The challenge of linkage attributes in EMS MTSs is primarily coordination among different components of the EMS delivery system. From the initial dispatch call to arrival at the ED, EMS personnel must always be cognizant of their role within the chain of emergency response. For the patient arriving in an ambulance to the ED, the paramedics would need to have contacted the ED to ensure that there was space for the patient and to prepare them for the patient's specific case. This includes conveying the severity of the situation as well as the current treatment and status of the patient to allow the ED to determine what resources (e.g., personnel, medication) will be required to be present upon arrival to treat the patient. This is especially crucial in patients with time-sensitive conditions such as heart attacks, stroke, trauma, and postresuscitation patients. While many EMS component teams tend to function sequentially, handing off the patient from one team to the next, patients who require longer-term care, such as cancer patients, have various teams coordinating their treatment. In these cases, certain teams have more responsibility and more influence in terms of patient treatment. For instance, an oncologist, because of expertise, would have more authority than a primary care physician or pharmacist in determining the overall care of a cancer patient even though the primary care physician or pharmacist may have the authority for specific elements of the patient's care. The final linkage attribute, communication methods and patterns, manifests in the ED when transferring information between teams or handing off a patient's care to another department or team, such as at the end of a shift. Information can be shared face to face, such as during handoffs, or electronically, as is more likely across organizational boundaries, each having different benefits and limitations for effective communication.

The development of an MTS in an ED relies on the case being presented to them. The medical personnel would be different if the patient has a broken arm compared to a heart attack. As such, most EMS MTSs emerge based upon what is needed and are not often appointed. Similarly, the duration of the MTS relies on the severity of the case and the types of treatment that the patient requires; treating a broken arm is significantly quicker than treating a patient who has had a heart attack. Finally, component teams and their connections to one another are quite fluid in EMS MTSs and can change dramatically and quickly depending on each patient's situation or can remain stable across time. If a patient arrives at the ED complaining of one ailment, the teams that work on treating that patient may change as new information arises, such as an underlying or worsening condition. For longer-term treatment plans, however, component teams are likely to remain relatively stable over time. Similarly, the connections among EMS teams will vary if the teams themselves change or shift depending on the patient's circumstances and needs.

A MASS CASUALTY EMS MTS EXAMPLE

The 911 dispatchers receive multiple calls of an explosion in or near the Federal Court Building with multiple people injured and some possibly killed. Due to the chaotic nature of the situation and the panicked callers, it is difficult for the 911 dispatchers using their established call-taking algorithms to determine accurate information about the number of injured individuals, the nature of those injuries, and the nature of the explosion. They are also unable to determine whether any of the callers have a sense of whether there are other explosive hazards on the scene. Following established protocols for these types of events, they send the maximal available response in the system from law enforcement, the fire department, and EMS. They also place a call to surrounding localities to alert them that they may be requesting their assistance as this event unfolds and additional information about the scope of the incident is obtained. In parallel, they notify the area's trauma centers and other hospitals that there has been an explosion at the Federal Court Building with an unknown number of injuries and that they will provide updates to the hospitals as more information becomes available. The largest medical center in the area activates their protocol as a command center to coordinate receiving and caring for the victims.

Law enforcement personnel are the first on the scene and immediately begin to secure the scene and search for additional threats to safety. Since the event appears to be a criminal activity with ongoing threat, they have authority for the scene. The first EMS providers arrive on the scene coordinated with law enforcement to ascertain that the scene is safe for them to enter and does not pose a risk to them or other arriving responders. They rely on and trust law enforcement for this information and to provide security to them and their patients. When authorized by law enforcement and directed by their understanding of the scene, the EMS personnel enter the taped-off scene and do a rapid survey of all victims to determine the number of injured and dead and to triage them by severity in anticipation of the arrival of additional EMS responders. They then begin to treat the most seriously wounded per protocol. They also advise their dispatch center to notify the receiving trauma centers and other hospitals of the incident and number of potential patients. Dispatchers coordinate with the command center at the largest hospital to identify which patients, based on the severity and type of injuries, should go to which hospital.

Meanwhile, additional police and fire resources arrive on the scene and begin to manage and direct traffic as well as assist the EMS providers. The fire department also sets up a secure landing zone for the incoming helicopters. As air rescue arrives, fire alerts EMS to this resource and helps distribute victims to the most suited method of travel.

Mass casualties such as those previously mentioned are rare but are a special challenge to the emergency response system. Compositionally, they require larger numbers of personnel by their very nature. In this example, the MTS does not just involve a unit from EMS, police, and fire but multiple units for each area. As such, it represents an extremely complex MTS. Effective coordination between the component teams is critical. To support these efforts developmentally, localities hold annual or more frequent mass casualty drills to simulate just such events. The majority of large- and mid-sized agencies utilize the National Incident Management System (NIMS) as the operating framework during drills and real-life incidents (reference: http://www .fema.gov/pdf/emergency/nims/NIMS_core.pdf). This system establishes a platform of command and control, communication, and logistics by which all involved agencies can interface and theoretically manage all threats and hazards regardless of size, complexity, or location. In order for this system to work effectively, individual team members and organizations have to be trained in how the system works and practice utilizing the NIMS framework. Each of the services has specific missions and tasks to accomplish. Even though these teams are working side by side, their missions may not always align. For example, a severely injured victim on a scene that has not been secured and has hazards that can injure EMS providers requires the highest level of communication to evaluate and treat the victim. Relying on effective communication, EMS responders who are not immediately allowed to enter a scene may provide guidance to the police or fire personnel who are with the victim. If the teamwork is not good, it can result in serious harm or death to the patient.

One of the key findings of most mass casualty drills is that communication is not effective. So many teams are working on the incident, many of whom have never met each other or worked together before. Some coordination issues are related to technical barriers to communication such as frequency differences in radios or other hardware. Communication barriers can also be related to protocol or training issues such as a poorly established or poorly understood chain of command. Training for and improving the response to mass casualty incidents require leadership commitment and resources across multiple organizations to identify barriers, improve response, and better prepare for the rare, significant event.

CHALLENGES OF MTS PROCESSES IN AN EMS CONTEXT

Although MTSs are a critical part of EMS, there are many challenges that can impact their effectiveness in this environment. First, the physical environment in which EMS providers operate can pose many challenges, particularly in comparison to more controlled in-patient locations (Misasi et al., 2014). These compositional settings typically involve ambiguous and potentially volatile environments, such as in a roadside automobile crash, whereby both time and physical challenges exist given the need to quickly assess, extricate, and relocate patients to a safer situation. EMS teams must coordinate with police, fire, and other groups. Additionally, these environments require EMS personnel to serve many functions simultaneously (e.g., nurse, pharmacist, physician), with limited resources available, forcing teams to work autonomously and to adapt quickly (Meisel et al., 2008; Nguyen, 2008). The EMS personnel are also the conduit to higher expertise, frequently initiating a cascade of more expert practitioners as they bring a patient to an ED. As such, while the composition of a specific EMS team is discrete, a primary function of this team is to function as an MTS and add additional teams of experts from the rest of the healthcare system as dictated by the patient's condition.

Given these demands, linkage or coordination at the system-level communication may often be complex. Crews may have very limited information regarding patient medical history, which can lead providers to develop and execute care plans that may quickly change as additional information is provided (New York University, 2005). Further, when component teams are not familiar with one another, they may focus on unimportant or distracting communication that may impede on the relay of critical information (DiazGranados et al., 2014; Hegner & Larson, 2014). However, it has also been noted that EMSs are often "left out of the loop" in regard to emergency preparedness training with other first responder agencies at both national (e.g., Department of Homeland Security) and local levels (e.g., firefighters, community-based emergency planning boards; Robyn, 2005). This lack of inclusion in training and preparation can create challenges during real emergencies. The EMS component teams within first responder MTSs may not have the appropriate communication protocols in place for sharing information with other teams, which may lead to patient errors (Weaver et al., 2014). Additionally, in terms of the developmental attribute without preparing as a system for responding to emergencies, there may be a lack of clear boundary spanners and liaisons within the system who serve to connect teams to one another and with outside agencies in order to ensure that appropriate care is being provided (Davison et al., 2012). Furthermore, by not preparing with others, first responders may feel that they do not identify as a part of the overall EMS MTS. This identification with the system can be critical in terms of impacting system-level processes and outcomes (Connaughton & Shuffler, 2007). Component teams that do not have a sense of belonging or identity with the larger system are more likely to experience conflict and trust issues and are less likely to work toward achieving goals for the good of the entire system, instead focusing on goals that are primarily relevant to their own team (Burke et al., 2014). Overall, there are numerous challenges that must be faced regarding how to effectively manage EMS MTSs.

IMPROVING MTS EFFECTIVENESS IN EMS RESPONSE

Overall, the challenges discussed thus far have important implications for improving the effectiveness of EMS MTSs. Coordination within and across MTS component teams has been identified as critical for MTS functioning (DeChurch & Marks, 2006). Particularly in an EMS MTS, the hypercomplexity and stakes in which this work is conducted make coordination of ultimate importance. Coordination, whether horizontal (i.e., with others of a similar position or level in the MTS) or vertical (i.e., with supervisors or managers), is difficult in this context given the division of responsibilities and diversity of expertise within the MTS. This diversity serves a functional goal; however, effective coordination, horizontally or vertically, permits teams and members of the MTS to better understand the functioning of the entire system, which may lead to increased knowledge of each other's roles, tasks, and responsibilities (Marks et al., 2002; Joshi et al., 2009).

The need for high levels of coordination creates implications for leadership as well. Leadership development is critical to improve the coordination of EMS MTS and its effectiveness. For example, it is necessary to ensure that the leaders and boundary spanners (who connect the component teams when necessary) of the component teams are trained to effectively coordinate, strategize, and orchestrate the efforts of the teams. Leaders of each component team must understand the demands and interdependencies related to the task and goals in their own teams and in the MTS as a whole. Moreover, the leaders involved in an EMS MTS must be able to understand how to develop an MTS-level action plan and provide the team members support in order to be effective.

Lastly, the challenges found in an EMS MTS, as highlighted in this chapter, have operational implications for how the component teams communicate and are trained.

In these types of MTSs, communication requires swift and concise communication for quick, algorithm-based decision-making, particularly in situations such as the mass casualty example. These communication skills are often developed during drills and practice simulations. However, training needs to address the development of skills and knowledge necessary to maintain performance levels so that decay does not occur, particularly given the rarity of emergency situations similar to mass casualty events.

CONCLUSION

The cases that emergency response professionals answer to are hypercomplex, with a large number of individuals and teams involved. This hypercomplexity requires the teams, which make up the MTS, to be tightly coupled. The multiple component teams (e.g., fire, police, EMS, hospitals) not only have unique team goals but also have shared MTS-level goals; they must balance when responding to emergency calls. As such, police, fire, and EMS professionals must have sufficient coordination across different missions, professions, and work structures to effectively and efficiently provide citizens with the expected services.

Research on team effectiveness within healthcare has primarily focused on the individual, with more recent research incorporating the team. The environment in which emergency professionals work is hypercomplex, and research needs to address this by investigating emergency responders in the context of an MTS. Given the paucity of research in the emergency response literature, more is needed to better understand the coordination among all of the component teams in an EMS MTS. There is also a need for research that examines the MTS in terms of compositional, linkage, and developmental characteristics as well as research focusing on the development of evidence-based prescriptions for improving MTS effectiveness. Similarly, research should explore the expertise and distinct capabilities of the component teams in the EMS MTS and how important integrative leadership is in this context (see de Vries et al., 2015). The hypercomplex environment in which emergency responders work is beyond a single individual or team's ability to handle. As such, basic research should focus on understanding the context and functioning of these MTSs, which can then inform organizational practices.

ACKNOWLEDGMENTS

This work was supported in part by the grant award UL1TR000058 funding received from the National Center for Advancing Translational Sciences (NCATS), a component of the National Institutes of Health (NIH), and the Greenville Health System (GHS).

REFERENCES

Burke, C.S., DiazGranados, D. & Heyne, K. (2014). Examining the role of trust in partially distributed multiteam systems. *Symposium Presentation: Society for Industrial-Organizational Psychology Annual Conference*, Honolulu, Hawaii.

Connaughton, S.L. & Shuffler, M. (2007). Multinational and multicultural distributed teams: A review and future agenda. *Small Group Research*, *38*(3), 387–412.

Davison, R.B., Hollenbeck, J.R., Barnes, C.M., Sleesman, D.J. & Ilgen, D.R. (2012). Coordinated action in multiteam systems. *Journal of Applied Psychology*, *97*, 808–824.

DeChurch, L.A., Burke, C.S., Shuffler, M.L., Lyons, R., Doty, D. & Salas, E. (2011). A historiometric analysis of leadership in mission critical multiteam environments. *The Leadership Quarterly*, *22*, 152–169.

DeChurch, L.A. & Mathieu, J.E. (2009). Thinking in terms of multiteam systems. In E. Salas, G. Goodwin & C. Burke (Eds.), *Team effectiveness in complex organizations: Cross-disciplinary perspectives and approaches* (pp. 267–292). New York: Taylor & Francis.

DeChurch, L.A. & Marks, M.A. (2006). Leadership in multiteam systems. *Journal of Applied Psychology*, *91*, 311.

de Vries, T.A., Hollenbeck, J.R., Davison, R.B., Walter, F. & Van der Vegt, G. (2015). Managing coordination in multiteam systems: Integrating micro and macro perspectives. *Academy of Management Journal*, *59*.

DiazGranados, D., Dow, A.W., Perry, S.J. & Palesis, J.A. (2014). Understanding patient care as a multiteam system. In M. Shuffler, R. Rico & E. Salas (Eds.), *Pushing the boundaries: Multiteam systems in research and practice* (pp. 95–113). Bingley, UK: Emerald Group Publishing.

Goodwin, G.F., Essens, P.J.M.D. & Smith, D. (2012). Multiteam systems in the public sector. In S. Zaccaro, M. Marks & L. DeChurch (Eds.), *Multiteam systems: An organization form for dynamic and complex environments* (pp. 53–80). New York: Taylor & Francis.

Hegner, R.E. & Larson, M. (2014). Multiteam systems in large-scale disaster recovery. In *Pushing the boundaries: Multiteam systems in research and practice* (pp. 77–91). Bingley, UK: Emerald Group Publishing.

Institute of Medicine. (2006). *Emergency medical services: At the crossroads.* Washington, DC: National Academies Press.

Joshi, A., Pandey, N. & Han, G.H. (2009). Bracketing team boundary spanning: An examination of task-based, team-level, and contextual antecedents. *Journal of Organizational Behavior*, *30*, 731–759.

Kobusingye, O.C., Hyder, A.A., Bishai, D., Joshipura, M., Hicks, E.R. & Mock, C. (2006). Emergency medical services. In D.T. Jamison, J.G. Breman, A.R. Measham, G. Alleyne, M. Claeson, D.B. Evans, A. Mills, P. Musgrove (Eds.), *Disease control priorities in developing countries* (pp. 1261–1279). Washington, DC: World Bank.

Marks, M.A., Sabella, M.J., Burke, C.S. & Zaccaro, S.J. (2002). The impact of crosstraining on team effectiveness. *Journal of Applied Psychology*, *87*, 3–13.

Mathieu, J.E. (2012). Reflections on the evolution of the multiteam systems concept and a look to the future. In S. Zaccaro, M. Marks & L. DeChurch (Eds.), *Multiteam systems: An organization form for dynamic and complex environments* (pp. 511–544). New York: Taylor & Francis.

Mathieu, J.E., Marks, M.A. & Zaccaro, S.J. (2001). Multi-team systems. In N. Anderson, D.S. Ones, H.K. Sinangil & C. Viswesvaran (Eds.), *Organizational psychology: Vol. 2. Handbook of industrial, work and organizational psychology* (pp. 289–313). London: Sage.

Meisel, Z.F., Hargarten, S. & Vernick, J. (2008). Addressing prehospital patient safety using the science of injury prevention and control. *Prehospital Emergency Care*, *12*(4), 411–416.

Misasi, P., Lazzara, E. & Keebler, J.R. (2014). Understanding multiteam systems in emergency care: One case at a time. In M. Shuffler, E. Salas & R. Rico (Eds.), *Pushing the boundaries: Multiteam systems in research and practice* (pp. 157–183). Bingley, UK: Emerald Group Publishing.

Moynihan, D.P. (2007). *From forest fires to Hurricane Katrina: Case studies of incident command systems.* Washington, DC: IBM Center for the Business of Government.

Nguyen, A. (2008). Bad medicine: Preventing drug errors in the prehospital setting. *JEMS: Journal of Emergency Medical Services*, *33*(10), 94–100.

New York University. (2005). Emergency medical services: The forgotten first responder. Retrieved from New York University, Center for Catastrophe Preparedness and Response website: http://www.nyu.edu/ccpr/NYUEMSreport.pdf.

Roberts, K.H. & Rousseau, D.M. (1989). Research in nearly failure-free, high-reliability organizations: Having the bubble. *IEEE Transactions on Engineering Management*, *36*, 132–139.

Robyn, K. (2005). The forgotten first responder. Retrieved from http://www.emsworld.com/article/10323876/the-forgotten-first-responder.

Shuffler, M.L., Rico, R. & Salas, E. (2014). Pushing the boundaries of multiteam systems in research and practice: An introduction. In M. Shuffler, R. Rico & E. Salas (Eds.), *Pushing the boundaries: Multiteam systems in research and practice* (pp. 3–16). Bingley, UK: Emerald Group Publishing.

Weaver, S.J., Che, X.X., Pronovost, P.J., Goeschel, C.A., Kosel, K.C. & Rosen, M.A. (2014). Improving patient safety and care quality: A multiteam system perspective. In M. Shuffler, R. Rico & E. Salas (Eds.), *Pushing the boundaries: Multiteam systems in research and practice* (pp. 35–60). Bingley, UK: Emerald Group Publishing.

Zaccaro, S.J., Marks, M.A. & DeChurch, L.A. (2012). Multiteam systems: An introduction. In S. Zaccaro, M. Marks & L. DeChurch (Eds.), *Multiteam systems: An organization form for dynamic and complex environments* (pp. 3–32). New York: Taylor & Francis.

9 Cognitive Aids in Emergency Medical Services

Keaton A. Fletcher and Wendy L. Bedwell

CONTENTS

If there were just one defining characteristic of EMS, it would unequivocally be flawless service as quickly as possible. As Kobusingye et al. (2006, p. 1261) suggest, EMS requires "*rapid* assessment, *timely* provision of appropriate interventions, and *prompt* transportation to the nearest appropriate health facility by the best possible means to *enhance* survival, *control* morbidity, and *prevent* disability" (emphasis added). Within this statement lies an obvious paradox: perform optimally *and* rapidly. Yet science and practice both suggest a clear speed–accuracy trade-off (Fitts, 1954). So consistent is this finding that it has been named Fitts's law (MacKenzie, 1992) and, thus, placed on equal footing with other indisputable phenomenon such as gravity and thermodynamics. But the presence of gravity did not stop humans from reaching the moon, nor have the laws of thermodynamics prevented discovery/ creation of a variety of superconductors. In short, every law is made to be broken— or at least bent. The question is, how can we bend Fitts's law to make EMS both exceptionally fast *and* highly accurate? We argue that the answer lies, at least in part, with cognitive aids.

Cognitive aids are tools designed to lessen the cognitive burden associated with completing tasks from collecting information to storing and retrieving it, ultimately making the individual more efficient (Rosenthal & Downs, 1985). For example, the *World Health Organization* created a *safety checklist* to increase efficiency and accuracy in surgeries, ensuring that all safety protocols are followed (Weiser et al., 2010). This checklist, and almost all cognitive aids, works by reducing the amount of information that an individual has to process or structure, by off-loading some of it to an external tool (Rosenthal & Downs, 1985). In doing so, cognitive aids reduce task demands and provide additional resources to the user, allowing for a level of efficiency that was previously unattainable. Furthermore, cognitive aids not only improve the demand–resource ratio but also change the way that a task is completed to enable greater precision (Rosenthal & Downs, 1985). The result is an increase in speed *and* accuracy.

To further elucidate how cognitive aids can benefit EMS, we first need to thoroughly define what we mean by cognitive aid. Thus, we discuss the scope and purpose of cognitive aids and then organize them into a general taxonomy. This will lead to the main focus of our effort, an integration of these two into a graphical heuristic, from which we will describe how each type should be trained and used to maximize both the efficiency and effectiveness of EMS.

SCOPE AND PURPOSE OF COGNITIVE AIDS

The purpose of any cognitive aid is to improve outcomes (e.g., response time, survival rate, employee strain), but the manner in which this is accomplished can differ. We suggest that cognitive aids vary along two dimensions: scope and purpose. First, focusing on scope, we argue that cognitive aids can directly target either cognition or behaviors. Although completely isolating behaviors and cognition is neither reasonable nor practical, we contend that cognitive aids can be designed to primarily target one over the other. Cognitive aids can affect outcomes by directly altering, simplifying, or eliminating behaviors (Rosenthal & Downs, 1985) *or* cognition (Block & Morwitz, 1999; Makany et al., 2009). This distinction is critical when considering the implementation of cognitive aids in EMS. Cognitive aids that primarily target behaviors (e.g., preloaded syringes, behavioral checklists) necessarily are limited in scope as they are designed to target one unique behavior or set of behaviors (Hales et al., 2008). This limited scope allows for increased efficacy within the specified context but decreased generalizability (e.g., a lumbar puncture checklist or EpiPens). However, behaviorally targeted aids tend to be relatively easy to implement and can result in rapid improvements in outcomes (e.g., Dunlap & Dunlap, 1989) as users need only to enact the specified behavior(s) to achieve the desired result(s). Aids that directly target cognition (e.g., mnemonics), however, can be more generalized in scope, allowing for increased utility across situations (Rosenthal & Downs, 1985). This, however, necessitates a sacrifice of detail (e.g., Situation, Background, Assessment, and Recommendation [SBAR]). Generally, these cognitive aids take longer to implement than behavioral aids but can have lasting effects due to relatively stable cognitive changes (Rosenthal & Downs, 1985).

Cognitive aids also vary in their purpose. They can be designed primarily to create structure or simplify processes. Cognitive aids that primarily target structure are designed to refine or create a scaffold, or prescriptive guide, upon which cognitive and behavioral processes are built, organized, and executed (Rosenthal & Downs, 1985). These aids can create lasting effects that generalize across providers but tend to be resource intensive at the outset, requiring significant effort to design and train (Scaife & Rogers, 1996; Hales et al., 2008). Alternatively, cognitive aids can be designed to simplify cognitive or behavioral processes. These aids are designed to eliminate certain task-based demands (Zhang, 1997); however, they tend to be much less generalizable by functioning in only one manner or targeting a specific set of behaviors or cognition (e.g., an abacus or flow chart; Szolovits & Pauker, 1978). This limits both the situations and/or the users that benefit from aid implementation (e.g., Szolovits & Pauker, 1978). Although they can be resource intensive in design, they are not necessarily in implementation, requiring less training than aids that aim to create structure (e.g., Szolovits & Pauker, 1978).

Combining these two dimensions (scope and purpose) provides four broad domains of cognitive aids (see Table 9.1), each of which has unique strengths and weaknesses. First, aids that structure cognition are designed to efficiently organize information for easy recall. For example, a mnemonic device (e.g., SBAR) provides a framework with which patient information can be analyzed, organized, and shared. Cognitive aids designed to simplify cognition, however, can target the initiation or execution of already organized information. Leaving a patient's chart open to a specific page enables quick retrieval of required information. Such aids mitigate the cognitive burden of tasks by off-loading it to an external aid. Cognitive aids that structure behavior, such as those that target cognition, are designed to create a framework that can guide efficient execution of behavior. Decision trees or ordered checklists, for example, provide users with the ideal order of steps to follow when executing a task. Lastly, cognitive aids designed to simplify behaviors remove a time-consuming or problematic aspect of the task in favor of off-loading it to an external tool. Preloaded

---

TABLE 9.1

Key Aspects of Scope and Purpose of Cognitive Aids

		Scope	
		Behaviors	Cognition
Purpose	Create structure	• Moderate scope • Rapid changes • Resource intensive to implement	• Wide scope • Permanent changes • Resource intensive to implement
	Simplify processes	• Limited scope • Rapid changes • Resource intensive to create	• Limited scope • Resource intensive to create

syringes, for example, simplify task execution by removing the time-consuming and error-prone step of dosing medication, allowing for quicker administration. Each of these four domains can be targeted by a variety of cognitive aids. In the following section, we outline six major types of cognitive aids, how they simplify or structure behaviors and cognition, and how they can be used in EMS.

TYPES OF COGNITIVE AIDS

CHECKLISTS

One of the most widely used cognitive aids is organized tools that outline criteria or steps, simplify concepts, and aid information recall (Hales et al., 2008). A checklist is a physical or digital list of steps, ideas, categories, or key points. They should be clear (e.g., simple language and easy-to-read font), in-depth (e.g., all necessary information is included), and, when describing a process of steps, must be in the proper order (Luten et al., 2002; Harrison et al., 2006; Hales et al., 2008). When combined with appropriate intervention techniques, checklists can help EMS avoid missing steps or failing to capture all necessary information (Catchpole et al., 2007). Checklists are primarily designed to structure behavior, ensuring that all aspects of a task are completed. In general, checklists are designed to give information regarding how to complete a task (i.e., behaviors) not what thoughts to have about a problem (i.e., cognition). Yet when checklists are expected to target cognition or to simplify behavior, rather than structure it, users can be left feeling frustrated. This frustration can lead to checklist fatigue—a loss of desire to use the aid (Hales et al., 2008). Despite flexibility and ubiquity, checklists are not *always* the answer in EMS. For example, when a patient is crashing, using a checklist to look at all vitals could waste valuable time in resuscitating the patient. In such case, an alarm on the device monitoring a particular system would be more appropriate.

ALARMS

Alarms are defined as auditory warnings that increase situational awareness and vigilance and can advise required actions (Catchpole et al., 2004). Alarms should signal severity (e.g., louder alarms for more serious problems) and location (e.g., tonal sweeps that allow for auditory localization) and be unique to each event type (Celi et al., 2001; Catchpole et al., 2004). They have the capability to alert individuals to the most important or pressing issues in an environment and are considered "hard defenses" against error (Reason, 2000). An alarm for cardiac arrest, for example, eliminates the need to repeatedly check heart rate and compare it to safe levels. Although alarms, like checklists, are designed to ultimately influence behavior, their primary purpose is structuring cognition. In other words, alarms do not provide much information about how to solve a problem (i.e., behaviors) but primarily inform individuals that there is an issue, and what that issue is (i.e., cognition). As previously mentioned, the pitch or pattern of alarm can trigger a cascade of thoughts that ultimately leads to actions designed to eliminate the cause of the alarm. This motivated behavior can occur faster and more accurately because individuals have been trained on the meaning of the alarms and have created an accurate and effective cognitive structure. The users then have little need to search their environments for the source, instead relying on their knowledge of the alarm to inform their thoughts and, ultimately, actions. However, like checklists, alarms have been subject to overuse and improper application, which leads to a desensitization to *all* alarms, a phenomenon named alarm fatigue (Cvach, 2012). Despite published standardization of medical device alarms (IEC, 2006), they are often difficult to discriminate and do a poor job of signaling the severity of the issue (Sanderson et al., 2006). Coupled with a lack of training on the nuances of the various alarms, this has increased rates of alarm fatigue and physiological strain (Morrison et al., 2003).

PHYSICAL TOOLS

Physical tools are unlike alarms and checklists in that they are designed to simplify behaviors and cognition, not structure them. Physical tools are tangible objects that are designed with the specific purpose of reducing human error or necessary effort through the simplification, restriction, or enhancement of behavior. This category of cognitive aid encompasses a vast array of objects, such as differently sized and colored tubing, preloaded syringes, or calculators, designed to reduce human error and simplify tasks. To be useful, physical tools should be cost-effective (e.g., no more expensive than the cost of the alternative and any lawsuits that result in errors associated with use of the alternative), easy to use (e.g., intuitive and ergonomically designed), and pilot tested by the intended users (Celi et al., 2001; Luten et al., 2002; Catchpole et al., 2007). For example, bar code technology, specifically designed packaging, and premixed medications are useful tools to prevent medication errors (Cohen et al., 2007)—a costly error in EMS. Use of these aids can reduce error by allowing for cognitive off-loading (Luten et al., 2002; Dror & Harnad, 2008). However, heavy reliance on the physical tool's shape or size can be problematic as there are few regulations governing standardization. Further, over-reliance on physical tools can cause complacency, creating new opportunities for errors, as was seen in the airline industry with the introduction of automatic pilot options (Parasuraman & Riley, 1997).

MNEMONIC DEVICES

Mnemonic devices are "cognitive cuing structures that typically are made up either of visual images or of words in the form of sentences or rhymes" (Bellezza, 1984, p. 252). These patterns are used to structure, represent, and aid in the recall of connected ideas. A commonly used mnemonic in healthcare, for example, is SBAR. This mnemonic device is used as a reminder of a clear and consistent manner in which information should be given from one practitioner to another during a patient handover (Philibert & Leach, 2005). Mnemonic devices are designed to primarily structure cognition. They organize information in a logical way, but mnemonic devices are entirely cognitive in nature and must be memorized and recalled when appropriate. This places more of a cognitive burden on the healthcare provider than some other forms of cognitive aids but has been shown to reduce cognitive burden when compared to unstructured information (Verhaeghen et al., 1992). Mnemonic devices should be easy to remember (e.g., short and/or evocative of vivid imagery) and, if describing a sequence of steps, be in order (Hales et al., 2008). Given that their power comes from structuring cognition, which may ultimately lead to structured behavior (e.g., check airway, breathing, circulation [ABC]), mnemonic devices designed to simplify thoughts or behaviors may be less effective. Furthermore, mnemonic devices are not standardized across contexts or even care providers. Thus, when EMS interact with workers at various institutions, the staff may not be familiar with the same mnemonic devices, leading to confusion and miscommunication. Returning to our handover example, although SBAR is the most common mnemonic, in a recent study, 24 different patient handoff mnemonics were identified by trained reviewers (Riesenberg et al., 2009). This could lead to the perpetuation of nonfunctional mnemonic devices, the disuse of those that are effective, or confusion over what information to convey in which manner. Ultimately, this reduces both the speed and the accuracy of care delivery, resulting in patient harm and provider stress.

CUES

Cues are digital or physical stimuli (e.g., posters, signs, or warnings) designed to elicit specific behavioral patterns. In other words, cues primarily structure behaviors. Cues have been shown to reinforce organizational culture (Pettigrew, 1979) and prevent or promote certain behaviors (e.g., Pittet et al., 2000; Ford & Torok, 2008). Ultimately, however, the purpose of any cue, whether it is a stop sign or a pop-up reminder that Windows has critical updates, is to initiate behavior that might not otherwise occur. In general, however, cues can be easily ignored. They may be effective at altering behavior upon initial implementation, but unless the perceived consequences of disobeying the cue are too high, the effects can rapidly diminish. Even though this pattern is seen with both physical and digital cues, digital cues (e.g., pop-up reminders on a computer or mobile device) can serve to initiate or prevent behaviors in a more real-time, situation-specific manner than their physical counterparts. This can help alleviate the problem of rapidly diminishing effects by creating novel stimuli or by limiting presentation of the cue to only relevant situations. In doing so, digital cues can structure behavior more effectively based on the situation at hand, perhaps without falling subject to inattentional blindness (Anderson et al., 2010). In EMS where time is of the essence, both physical and digital cues hold promise due to their ability to rapidly structure behavior, reducing the time required to act, and increasing accuracy, all without placing a greater burden on the worker.

DECISION SUPPORT SYSTEMS

Decision support systems (DSSs) are defined in this chapter as algorithm-based tools that take information regarding a situation or patient and provide suggestions regarding courses of action (Miller et al., 2015). These can be as simple as paper-based flow charts or as complex as programs integrated into the electronic medical/health records. DSSs rely upon empirically derived algorithms to lead the user to the most likely diagnosis or the best course of action. In essence, this is the same process that a human goes through when mentally making decisions, but we are subject to a range of biases and errors. DSSs are designed to simplify this process, reducing the possibility for human error. Certainly, DSSs are only as effective as the algorithms from whence they are derived and the inputs with which they create output, but certainly, they simplify cognition. In fact, a well-designed cognitive aid should not only increase accuracy but should also increase the speed with which the user can come to a conclusion regarding the situation and react. As such, DSSs can be incredibly useful for EMS. Digital DSSs, in particular, hold great promise. Many DSSs are able to eliminate the need for human input, instead sampling the environment or patient and providing recommendations. Blood pressure monitors, for example, which automatically provide the user with a suggested diagnosis of hypertension, prehypertension, or arrhythmia, can save the user time and can be more accurate, in trying to determine if the blood pressure or heart rate is abnormal. Furthermore, modern technology has increased the amount of accessible information, the speed with which it can be used, and the vast number of DSSs that can tap into this information (DesRoches et al., 2008). Mobile applications, for example, have allowed a tablet or cell phone to rapidly provide EMS with suggested diagnoses and/or course of action, all without sacrificing time. However, similar to the aforementioned cognitive aid types, over-reliance on these aids has resulted in dissatisfaction from both the care provider and the patient.

A VISUAL HEURISTIC OF COGNITIVE AIDS

Combining the six types of cognitive aids (Table 9.2) with the two design dimensions (scope and purpose), we have created a visual heuristic (Figure 9.1) to help determine which cognitive aids should be used to target specific goals in EMS, how EMS should be trained to use them, and how the aids can be paired to address unique problems in EMS.

Cognitive aids in quadrant 1 are designed to structure behaviors and include checklists and cues. The cognitive aids in quadrant I are not meant to provide novel information but rather are used to create structure for behaviors that have already been learned. *SPEEDBOMB* (Mommers & Keogh, 2015), for example, is a checklist designed to confirm that the user has properly executed preintubation steps (e.g., *suction, positioning, equipment*). This checklist is not designed to teach these tasks but can help to ensure that the executed behaviors follow the proper sequence and are completed properly. Similarly, *SleepTrackTXT* (Patterson et al., 2014), which is a digital cue system that helps EMS to track their fatigue levels, is not designed to inform providers that fatigue is dangerous. It is not even designed to teach providers how to avoid fatigue but rather to elicit a specific structure of behaviors that have already been trained.

Cognitive aids located in quadrant II, designed to structure cognition, include mnemonics and alarms. For example, *RASTAFARI* (Legrand et al., 2015) is a mnemonic designed to help providers learn and properly sequence the steps of prehospital and early in-hospital management (e.g., *rule of 9, associated trauma and intoxication, secure airway*). This mnemonic is helpful not only when learning this sequence by simplifying and structuring the information into easily remembered chunks, but

TABLE 9.2

Taxonomy of Cognitive Aids and Examples in EMS

Aid	Definition	Examples in EMS
Checklists	Organized tools that outline criteria or steps, simplify concepts, and aid information recall (Hales et al., 2008)	• SPEEDBOMB (Mommers & Keogh, 2015) • General, Acute Coronary Syndrome, Asthma and chronic obstructive pulmonary disease checklists (Kerner et al., 2015)
Alarms	Warnings that increase situational awareness and vigilance, and can advise required actions (Catchpole et al., 2004)	• MWVSM (Van Haren et al., 2014) • Ventricular-assist devices alarms (Mechem, 2013)
Physical tools	Tangible objects designed to reduce human error or necessary effort through the simplification, restriction, or enhancement of behavior	• Preloaded bougie (Baker et al., 2015) • LUCAS-2 (Perkins et al., 2015)
Mnemonic devices	"Cognitive cuing structures that typically are made up either of visual images or of words in the form of sentences or rhymes" (Bellezza, 1984, p. 252)	• RASTAFARI (Legrand et al., 2015) • I-PASS (Starmer et al., 2012)
Cues	Digital or physical stimuli (e.g., posters, signs, reminders, or warnings) designed to elicit specific behavioral patterns	• SleepTrackTXT (Patterson et al., 2014) • Universal Barrier Precautions posters (Kelen et al., 1990)
DSSs	Algorithm-based tools that use situation or patient information to provide potential courses of action (Miller et al., 2015)	• DSS for prehospital care (Vicente et al., 2013) • Acute cardiac ischemia TIPI (Selker et al., 1991)

also during task execution, by continuing to structure the cognitive processes of EMS by providing a framework upon which the situation-specific information can be interpreted. Another example of a cognitive aid in quadrant II is the alarm system associated with the *miniature wireless vital signs monitor* (*MWVSM*; Van Haren et al., 2014). The alarm sounds when the patient's vital signs fall within one of five predetermined status regions. These alarms elicit a set of associated cognition that has been learned (through either formal training or informal learning). The alarm can be helpful when first learning the sets of risky vital sign combinations by providing a set structure upon which the new information can be built but continues to be helpful by eliciting these learned categories.

DSSs are located primarily in quadrant III, as they are designed to simplify cognition. The *acute cardiac ischemia time-insensitive predictive instrument* (*TIPI*; Selker et al., 1991) is a simple algorithm that combines seven variables that can be collected upon initial contact with the patient and provides a probability of acute cardiac ischemia. This decision aid can be incorporated into a digital program or hand calculated; either way, it simplifies the decision process of whether a patient should

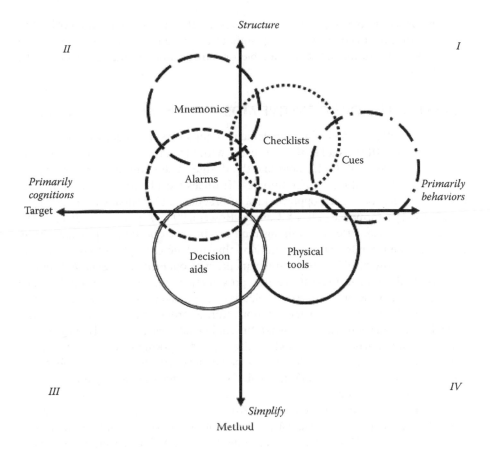

FIGURE 9.1 A visual heuristic of cognitive aids and their methods and targets. Overlapping circles represent potentially useful pairings of cognitive aids.

be admitted or triaged. EMS using this DSS need not know or be trained on the math or research that led to the development of the TIPI, nor could they be more effective without this cognitive aid. As such, the TIPI should and many other decision aids simplify cognition and allow EMS to rest upon the work and research of many others in order to rapidly deliver the best and most appropriate care.

Physical tools are primarily located in quadrant IV because they focus primarily on simplifying behaviors. The preloaded bougie (Baker et al., 2015), for example, is designed to optimize the process of endotracheal tube (ETT) placement and is one of many similar ETT introducers. Bougies are a general tool designed to supplement visualization or to overcome patient factors that make ETT introduction difficult by providing a guided path upon which the ETT can be sent into place. For EMS, this multistep process has been simplified through the design of preloaded bougies, upon which the ETT is already placed. This not only reduces the number of steps required to complete the task, which could save critically important seconds in an emergency situation, but can also reduce the chance of human error and provider stress under the pressures of said emergency situation. Despite the utility of physical tools, EMS should know how to complete the task without the benefits of simplification so that in situations where the aid is unavailable or malfunctioning, the task can still be completed.

TRAINING USE OF COGNITIVE AIDS

Maximizing effectiveness of any implemented cognitive aids requires proper training (Marshall, 2013), as it familiarizes users with the aids' purpose, method of use, and utility. Salas et al. (2012) note that the design of the training can improve (or hinder) knowledge, skills, and/or attitudes. Trainees must be provided with instruction, demonstration, varied opportunities to practice, and feedback regarding their practice and performance (Salas & Cannon-Bowers, 2001; Aguinis & Kraiger, 2009; Salas et al., 2012). Effective training requires information that is provided in an engaging manner, relevant demonstrations that include examples of good and bad behaviors (Baldwin, 1992), and job-relevant practice opportunities (Salas & Cannon-Bowers, 2001). Specific to cognitive aids, presentation of information should include the problem that the aid addresses (typically the speed–accuracy trade-off), appropriate uses, and limitations. Demonstration requires proper use of the aid and the targeted behavior in a job-relevant context. Practicing aid use and learned behaviors can occur in simulation or on the job and requires varied opportunities to explore various methods of interaction. Regardless of practice context, timely and relevant feedback must be provided by an expert (Salas et al., 2008). Trainers should explain the utility of the cognitive aid as it relates to EMS to enhance transfer of skills (Alliger et al., 1997). This information can be garnered from an appropriate training need analysis that takes into consideration the purpose and ultimate goals of the training session (Salas et al., 2012). With regard to cognitive aids in EMS, consideration of which quadrant houses the targeted cognitive aid can facilitate this process (see Table 9.3).

Training Quadrant I

Focusing on quadrant I, aids primarily designed to structure behavior have specific needs with regard to practice. Information provision should focus on detailing the utility of the aid, when it is and is not appropriate, and how it fits into current cognitive structures (e.g., other aspects of a larger training program). Demonstration should include both proper and improper uses (Baldwin, 1992), given that these aids tend to be fairly flexible but have only one best method of use. Unlike many other cognitive aids, however, those that lie primarily in quadrant I are designed for specific circumstances, and thus, practice should occur within these situations or should target these specific situations alone. It has been shown that varied practice yields better results than simply overlearning a skill in one situation (Machin & Fogarty, 2003), but given that there is only one (or perhaps a handful) of situations in which this cognitive aid will be used, dramatically varying unnecessary uses is rather fruitless. Instead, varied practice should focus more on adaptability and how to proceed to use the aid properly if certain aspects of the task or environment proceed unexpectedly.

TABLE 9.3

Training Cognitive Aid Use by Quadrant

	Information	Demonstration	Practice	Feedback
Quad I	• Necessity of the aid • How it fits into cognitive structure	• Proper and improper uses	• Varied conditions • Only target relevant situations	• Focus on promoting use of these aids • What went well and what did not
Quad II	• Distributed overlearning	• Walk-through cognition • Less necessary	• Distributed opportunities to recall learned structure	• Focus on promoting use of these aids • What went well and what did not
Quad III	• Necessity of the aid • Do not overcomplicate	• Proper and improper uses • Task without use of the aid	• Varied situations and conditions	• Focus on promoting use of these aids • What went well and what did not
Quad IV	• What behaviors it is simplifying • Why and how it is should be used	• Proper and improper uses • Task without use of the aid	• Varied situations and conditions	• Focus on promoting use of these aids • What went well and what did not

For checklists, the practice aspect of training should focus on adhering to each step or acknowledging each item on the checklist regardless of the situation. For example, the *Necessary, Enough, Working, and Secure* (Schoettker et al., 2003) checklist was designed for advanced trauma life support, providing workers with a reminder to check and record problems with the airway, breathing, circulation, etc. This checklist is typically used for patients who experience road trauma, falls, or violence. Each patient should receive the same overall checks, but the differences between each type of patient and their situation necessitate adaptation in some capacity. Adaptability, in this sense, can certainly be trained and prepared through practicing in varied contexts. This ensures that the procedure (i.e., structure of the behaviors) remains consistent regardless of dynamic, unique, or challenging situations.

Cues similarly require varied practice to be maximally effective. The efficacy of a cue is dictated not only by the degree to which the appropriate behavior is learned through training but also by the ability for the reminder to activate behaviors in the moment. Take the *level 1 MI protocol* (Henry et al., 2005), for example, that provides guidance regarding how to appropriately stabilize and transfer patients with potential myocardial infarction to the *Minneapolis Heart Institute*. This protocol is printed on posters in emergency departments of all local hospitals, as well as on index cards that are distributed to area EMS. Individuals need to be trained not only to respond to the rare occurrence of myocardial infarction through the effective and accurate use of this cue but also to recognize situations in which this cue is relevant. Only through varied practice (situations and patients) can EMS learn the latter, while the former should proceed according to the cue and trained behavior regardless of the situation.

TRAINING QUADRANT II

Shifting focus to quadrant II, which houses all cognitive aids that seek to structure cognition, we find that training these aids requires overlearning of information. Given that the purpose of this type of cognitive aid is to change the manner in which the user thinks, it makes sense that training the use of these aids should primarily focus on long-term cognitive changes. The best way in which a training program can permanently affect cognitive structure is through distributed overlearning (Rohrer & Pashler, 2007). Further, given that behavioral changes are not the primary, but rather secondary, target for these aids, a heavy focus on demonstration and practice is not necessarily warranted, nor practical. Demonstrations that do occur should include walk-throughs of elicited cognition, so as to help create cognitive structures that can be refined at a later date. Alarms, for example, are only as effective as the cognition that they elicit, and unidentified alarms are problematic. Training should focus on alarm exposure to ensure that EMS can identify each type. Alarms that have been undertrained can be a distraction, forcing the provider to determine the cause of the alarm instead of focusing on the task. Alarms that have been overtrained can elicit a cognitive structure immediately, without significantly distracting the care provider, instead automatically initiating a response. Take the different alarm patterns of a pager-based alarm system such as that proposed by Russek (1994) and that of the user's mobile phone. Both are silent and rely upon vibrations, but through repeated exposure, the user has overlearned which vibration patterns indicate that a patient needs emergency medical attention and the other indicates the arrival of a text message.

Mnemonics, too, must be overlearned through distributed rehearsal. The extent to which a mnemonic is effective depends on the ease with which it is used. Many mnemonics exist to target a variety of cognitions (e.g., SBAR, *Introduction, Situation, Background, Assessment, and Recommendation* [ISBAR], GRIEVING), but each mnemonic is designed to simply provide the user with a structure upon which cognition can be organized. These mnemonics need to be overlearned; otherwise, the user will exert effort trying to remember the mnemonic, which is only a mechanism, not the outcome. For example, when transitioning a patient to new care, the EMS provider should automatically be prepared to follow SBAR or ISBAR rather than trying to remember what mnemonic to use and what each letter represents. Instead, the mnemonic should structure how cognitions are activated, organized, and communicated. This can be achieved only through overlearning, repeated exposure to the mnemonic, and extensive rehearsal. There are many mnemonics that an EMS provider must memorize, but each is useful in its own settings, but even the rarely used mnemonics need to be readily activated.

TRAINING QUADRANT III

Cognitive aids in quadrant III are designed to simplify cognition. In order to do so, it is imperative that users are given ample opportunities to practice using the aid.

Unlike cognitive aids in quadrant I, those in quadrant III should be trained in a much more varied manner, allowing the user to explore the aid and how it can benefit him/her. It is difficult to dictate how a cognitive aid should simplify the user's cognition, given the variability between individuals in cognitive structures. However, if the user is given ample opportunities to explore how the cognitive aid can most effectively supplement the current cognitive structure, perhaps by bypassing certain steps, or entire processes, then the aid becomes exponentially more useful to the individual user. Further, information regarding the aid should include simply how the aid can be helpful and reduce the burden of the task, as well as how to properly use the aid. There is no need to explain how the aid works given the purpose of simplifying cognition, not making them more complex. Demonstration should include proper and improper uses of the aid and potentially what task execution would look like without the aid to impart perceptions of utility.

The primary cognitive aid type in quadrant III is DSSs. This type of cognitive aid typically converts a large amount of input into a single output, but the number of potential outputs is often vast. Flow charts, for example, rarely lead to one conclusion but rather help guide the user through a series of hurdles to one of many potential outcomes. Only through varied exploratory practice can individuals truly grasp the potential uses of the flow chart and how it can supplement or replace certain aspects of their cognitive processes. Digital DSSs are even more reliant upon varied exploratory practice given their seemingly limitless outputs. If users are not comfortable with the interface, something that can be trained through varied experiential learning and practice (De Freitas & Oliver, 2006), then they will be less likely to use the DSS, and thus, it loses its efficacy. Furthermore, through varied experiential practice and feedback, with the DSSs, the user may develop a generalized sense of efficacy with the aid, thus increasing the likelihood of use in a variety of situations.

TRAINING QUADRANT IV

Cognitive aids located in quadrant IV are targeted at simplifying behaviors. Information provided about the aid should include which behaviors it is simplifying, why, and how. This should help increase buy-in as well as help to create a mental model regarding how the aid can be used on the job. It is also imperative that information and feedback clarify proper and improper uses. However, it has been shown that for behaviors in which there is only one correct method of execution (e.g., using an EpiPen), it is more effective to show only the proper technique and to not demonstrate improper use. Practice should be accompanied with timely feedback that again clarifies what practiced behaviors were correct and in-line with the proper use of the cognitive aid and which behaviors need to be altered in order to better use the aid.

Physical tools are the primary type of cognitive aid in quadrant IV and as such should receive specialized training focusing on proper use. Whether the physical tool is a breathing bag, a preloaded syringe, or a calculator, it is imperative that the user knows exactly how to use it. Almost always, there is only one proper way to use a physical tool. Certainly, context may dictate variations in this behavior (e.g., varying placement of the electrodes of an automated external defibrillator on an infant, versus a child, versus an adult), but there is still only one correct way to use the aid.

By demonstrating proper use and providing feedback on the EMS provider's usage of the aid, training can help ensure accurate and speedy usage and, ultimately, delivery of care.

FINAL CONSIDERATIONS FOR COGNITIVE AIDS IN EMS

Practice and theory alike have suggested that various cognitive aids should be helpful in providing EMS with a way in which to increase both speed and accuracy. Yet consistent findings regarding the efficacy of cognitive aids in EMS remain elusive. Unlike many other high-performance fields such as aviation and nuclear safety, EMS includes a diverse population of practitioners, environments, and tasks. It is, therefore, necessary to abandon the notion of "one size fits all" when it comes to cognitive aids for EMS and to understand how to address unique problems in the field. This chapter provides a brief overview of the plethora of cognitive aid types as well as a visual heuristic with which EMS and managers can begin to discuss cognitive aids in a systematic way. Future research needs to look at the incremental effects of pairing cognitive aids (e.g., combined alarms and physical tools), as well as any potential trade-offs that may occur when using multiple cognitive aids at once.

Using the previous information, EMS and managers should be better able to implement cognitive aids or pairs of cognitive aids to address certain work tasks, depending on the unique demands. Further, by better understanding the scope and purpose of each cognitive aid, EMS managers can better provide training to ensure that cognitive aid-based interventions are maximally effective. Ultimately, we suggest that cognitive aids provide a unique and underutilized solution for EMS, facing overwhelming demands on their time and abilities.

REFERENCES

Aguinis, H. & Kraiger, K. (2009). Benefits of training and development for individuals and teams, organizations, and society. *Annual Review of Psychology, 60,* 451–474.

Alliger, G.M., Tannenbaum, S.I., Bennett Jr., W., Traver, H. & Shotland, A. (1997). A meta-analysis of the relations among training criteria. *Personnel Psychology, 50,* 341–358.

Anderson, J., Gosbee, L.L., Bessesen, M. & Williams, L. (2010). Using human factors engineering to improve the effectiveness of infection prevention and control. *Critical Care Medicine, 38,* S269–S281.

Baker, J.B., Maskell, K.F., Matlock, A.G., Walsh, R.M. & Skinner, C.G. (2015). Comparison of preloaded bougie versus standard bougie technique for endotracheal intubation in a cadaveric model. *Western Journal of Emergency Medicine, 16,* 588–593.

Baldwin, M.W. (1992). Relational schemas and the processing of social information. *Psychological Bulletin, 112,* 461–484.

Bellezza, F.S. (1984). The self as a mnemonic device: The role of internal cues. *Journal of Personality and Social Psychology, 47,* 506–516.

Block, L.G. & Morwitz, V.G. (1999). Shopping lists as an external memory aid for grocery shopping: Influences on list writing and list fulfillment. *Journal of Consumer Psychology, 8,* 343–375.

Catchpole, K.R., De Leval, M.R., Mcewan, A., Pigott, N., Elliot, M.J., Mcquillan, A. et al. (2007). Patient handover from surgery to intensive care: Using formula 1 pit-stop and aviation models to improve safety and quality. *Pediatric Anesthesia, 17,* 470–478.

Catchpole, K.R., McKeown, J.D. & Withington, D.J. (2004). Localizable auditory warning pulses. *Ergonomics, 47*, 748–771.

Celi, L.A., Hassan, E., Marquardt, C., Breslow, M. & Rosenfeld, B. (2001). The eICU: It's not just telemedicine. *Critical Care Medicine, 29*, N183–N189.

Cohen, M.R., Smetzer, J.L., Tuohy, N.R. & Kilo, C.M. (2007). High-alert medications: Safeguarding against errors. In M.R. Cohen (Ed.), *Medication errors.* 2nd ed. (pp. 317–411). Washington, DC: American Pharmaceutical Association.

Cvach, M. (2012). Monitor alarm fatigue: An integrative review. *Biomedical Instrumentation & Technology, 46*, 268–277.

De Freitas, S. & Oliver, M. (2006). How can exploratory learning with games and simulations within the curriculum be most effectively evaluated? *Computers & Education, 46*, 249–264.

DesRoches, C.M., Campbell, E.G., Rao, S.R., Donelan, K., Ferris, T.G., Jha, A. et al. (2008). Electronic health records in ambulatory care: A national survey of physicians. *New England Journal of Medicine, 359*, 50–60.

Dror, I.E. & Harnad, S. (2008). Offloading cognition onto cognitive technology. In I.E. Dror & S. Harnad (Eds.), *Cognition distributed: How cognitive technology extends our minds* (pp. 1–23). Philadelphia, PA: John Benjamins Publishing.

Dunlap, L.K. & Dunlap, G. (1989). A self-monitoring package for teaching subtraction with regrouping to students with learning disabilities. *Journal of Applied Behavior Analysis, 22*, 309–314.

Fitts, P.M. (1954). The information capacity of the human motor system in controlling the amplitude of movement. *Journal of Experimental Psychology, 47*, 381–391.

Ford, M.A. & Torok, D. (2008). Motivational signage increases physical activity on a college campus. *Journal of American College Health, 57*, 242–244.

Hales, B., Terblanche, M., Fowler, R. & Sibbald, W. (2008). Development of medical checklists for improved quality of patient care. *International Journal for Quality in Health Care, 20*, 22–30.

Harrison, T.K., Manser, T., Howard, S.K & Gaba, D.M. (2006). Use of cognitive aids in a simulated anesthetic crisis. *Anesthesia & Analgesia, 103*, 551–556.

Henry, T.D., Unger, B.T., Sharkey, S.W., Lips, D.L., Pedersen, W.R., Madison, J.D. et al. (2005). Design of a standardized system for transfer of patients with ST-elevation myocardial infarction for percutaneous coronary intervention. *American Heart Journal, 150*, 373–384.

International Electrotechnical Commission. (2006). *General requirements for basic safety and essential performance—Collateral standard: General requirements, tests and guidance for alarm systems in medical electrical equipment and medical electrical systems* (International Standard IEC 60601-8-6). International Electrotechnical Commission.

Kelen, G.D., DiGiovanna, T.A., Celentano, D.D., Kalainov, D., Bisson, L., Junkins, E. et al. (1990). Adherence to universal (barrier) precautions during interventions on critically ill and injured emergency department patients. *Journal of Acquired Immune Deficiency Syndromes, 3*, 987–994.

Kerner, T., Schmidbauer, W., Tietz, M., Marung, H. & Genzwuerker, H.V. (2015). Use of checklists improves the quality and safety of prehospital emergency care. *European Journal of Emergency Medicine: Official Journal of the European Society for Emergency Medicine*. Advance online publication. doi: 10.1097/MEJ.0000000000000315

Kobusingye, O.C., Hyder, A.A., Bishai, D., Joshipura, M., Hicks, E.R. & Mock, C. (2006). Emergency medical services. In D.T. Jamison, J.G. Breman, A.R. Measham, G. Alletne, M. Claeson, D.B. Evans et al. (Eds.), *Disease control priorities in developing countries.* 2nd ed. (pp. 1261–1279). Washington, DC: World Bank.

Legrand, M., Guttormsen, A.B. & Berger, M.M. (2015). Ten tips for managing critically ill burn patients: Follow the RASTAFARI! *Intensive Care Medicine, 41*, 1107–1109.

Luten, R., Wears, R.L., Broselow, J., Croskerry, P., Joseph, M.M. & Frush, K. (2002). Managing the unique size-related issues of pediatric resuscitation: Reducing cognitive load with resuscitation aids. *Academic Emergency Medicine, 9*, 840–847.

Machin, M.A. & Fogarty, G.J. (2003). Perceptions of training-related factors and personal variables as predictors of transfer implementation intentions. *Journal of Business and Psychology, 18*, 51–71.

MacKenzie, I.S. (1992). Fitts' law as a research and design tool in human-computer interaction. *Human-Computer Interaction, 7*, 91–139.

Makany, T., Kemp, J. & Dror, I.E. (2009). Optimising the use of note-taking as an external cognitive aid for increasing learning. *British Journal of Educational Technology, 40*(4), 619–635.

Marshall, S. (2013). The use of cognitive aids during emergencies in anesthesia: A review of the literature. *Anesthesia & Analgesia, 117*, 1162–1171.

Mechem, C.C. (2013). Prehospital assessment and management of patients with ventricular-assist devices. *Prehospital Emergency Care, 17*, 223–229.

Miller, A., Moon, B., Anders, S., Walden, R., Brown, S. & Montella, D. (2015). Integrating computerized clinical decision support systems into clinical work: A meta-synthesis of qualitative research. *International Journal of Medical Informatics, 84*, 1009–1018.

Mommers, L. & Keogh, S. (2015). SPEEDBOMB: A simple and rapid checklist for prehospital rapid sequence induction. *Emergency Medicine Australasia, 27*, 165–168.

Morrison, W.E., Haas, E.C., Shaffner, D.H., Garrett, E.S. & Fackler, J.C. (2003). Noise, stress, and annoyance in a pediatric intensive care unit. *Critical Care Medicine, 31*, 113–119.

Parasuraman, R. & Riley, V. (1997). Humans and automation: Use, misuse, disuse, and abuse. *Human Factors: The Journal of the Human Factors and Ergonomics Society, 39*(2), 230–253.

Patterson, P.D., Moore, C.G., Weaver, M.D., Buysse, D.J., Suffoletto, B.P., Callaway, C.W. et al. (2014). Mobile phone text messaging intervention to improve alertness and reduce sleepiness and fatigue during shiftwork among emergency medicine clinicians: Study protocol for the SleepTrackTXT pilot randomized controlled trial. *Trials, 15*, 1–10.

Perkins, G.D., Lall, R., Quinn, T., Deakin, C.D., Cooke, M.W., Horton, J. et al. (2015). Mechanical versus manual chest compression for out-of-hospital cardiac arrest (PARAMEDIC): A pragmatic, cluster randomised controlled trial. *Lancet, 385*, 947–955.

Pettigrew, A.M. (1979). On studying organizational cultures. *Administrative Science Quarterly, 24*, 570–581.

Philibert, I. & Leach, D.C. (2005). Re-framing continuity of care for this century. *Quality and Safety in Health Care, 14*, 394–396.

Pittet, D., Hugonnet, S., Harbarth, S., Mourouga, P., Saucan, V., Touveneau, S. et al. (2000). Effectiveness of a hospital-wide programme to improve compliance with hand hygiene. *Lancet, 356*, 1307–1312.

Reason, J. (2000). Human error: Models and management. *British Medical Journal, 320*, 768–770.

Riesenberg, L.A., Leitzsch, J. & Little, B.W. (2009). Systematic review of handoff mnemonics literature. *American Journal of Medical Quality, 24*, 196–204.

Rohrer, D. & Pashler, H. (2007). Increasing retention without increasing study time. *Current Directions in Psychological Science, 16*, 183–186.

Rosenthal, T.L. & Downs, A. (1985). Cognitive aids in teaching and treating. *Advances in Behaviour Research and Therapy, 7*, 1–53.

Russek, L.G. (1994). *US Patent No. 5,319,355*. Washington, DC: US Patent and Trademark Office.

Salas, E., Klein, C., King, H., Salisbury, M., Augenstein, J.S., Birnbach, D.J., Robinson, D.W. & Upshaw, C. (2008). Debriefing medical teams: 12 evidence-based best practices and tips. *The Joint Commission Journal on Quality and Patient Safety, 34*(9), 518–527.

Salas, E. & Cannon-Bowers, J.A. (2001). The science of training: A decade of progress. *Annual Review of Psychology, 52,* 471–499.

Salas, E., Tannenbaum, S.I., Kraiger, K. & Smith-Jentsch, K.A. (2012). The science of training and development in organizations: What matters in practice. *Psychological Science in the Public Interest, 13,* 74–101.

Sanderson, P.M., Wee, A. & Lacherez, P. (2006). Learnability and discriminability of melodic medical equipment alarms. *Anaesthesia, 61,* 142–147.

Scaife, M. & Rogers, Y. (1996). External cognition: How do graphical representations work? *International Journal of Human-Computer Studies, 45,* 185–213.

Schoettker, P., D'Amours, S.K., Nocera, N., Caldwell, E. & Sugrue, M. (2003). Reduction of time to definitive care in trauma patients: Effectiveness of a new checklist system. *Injury, 34,* 187–190.

Selker, H.P., Griffith, J.L. & D'Agostino, R.B. (1991). A tool for judging coronary care unit admission appropriateness, valid for both real-time and retrospective use: A time-insensitive predictive instrument (TIPI) for acute cardiac ischemia: A multicenter study. *Medical Care, 29,* 610–627.

Starmer, A.J., Spector, N.D., Srivastava, R., Allen, A.D., Landrigan, C.P. & Sectish, T.C. (2012). I-pass, a mnemonic to standardize verbal handoffs. *Pediatrics, 129,* 201–204.

Szolovits, P. & Pauker, S.G. (1978). Categorical and probabilistic reasoning in medical diagnosis. *Artificial Intelligence, 11,* 115–144.

Van Haren, R.M., Thorson, C.M., Valle, E.J., Busko, A.M., Jouria, J.M., Livingstone, A.S. et al. (2014). Novel prehospital monitor with injury acuity alarm to identify trauma patients who require lifesaving intervention. *Journal of Trauma and Acute Care Surgery, 76,* 743–749.

Verhaeghen, P., Marcoen, A. & Goossens, L. (1992). Improving memory performance in the aged through mnemonic training: A meta-analytic study. *Psychology and Aging, 7,* 242–251.

Vicente, V., Sjöstrand, F., Sundström, B.W., Svensson, L. & Castren, M. (2013). Developing a decision support system for geriatric patients in prehospital care. *European Journal of Emergency Medicine, 20,* 240–247.

Weiser, T.G., Haynes, A.B., Dziekan, G., Berry, W.R., Lipsitz, S.R. & Gawande, A.A. (2010). Effect of a 19-item surgical safety checklist during urgent operations in a global patient population. *Annals of Surgery, 251,* 976–980.

Zhang, J. (1997). The nature of external representations in problem solving. *Cognitive Science, 21,* 179–217.

10 Exploring Telemedicine in Emergency Medical Services
Guidance in Implementation for Practitioners

Elizabeth H. Lazzara and Lauren E. Benishek

CONTENTS

INTRODUCTION

Due to the growing specialties, evolving technologies, and disease complexities, quality care is now contingent upon the synthesis of expertise and seamless coordination of multiple, oftentimes distributed, individuals. EMS exemplify the aforementioned depiction, as it involves crew members serving multiple functions separately (e.g., managing patients and driving ambulances) as well as technology enabling the transmission of information between crews in the field and ED clinicians to foster patient management and preparation for incoming patient admissions.

To integrate seemingly disparate team members and their expertise while addressing patient needs and navigating the challenging clinical care system, the healthcare community has leveraged telemedicine. Currently, half of all US hospitals already use some form of telemedicine according to the American Telemedicine Association, and telemedicine is projected to expand significantly with the number of patients using telemedical services rising to seven million (Roashan, 2014). Also demonstrating this proliferation, previously valued at $14.2 billion, telemedicine is projected to swell at a compound annual growth rate of 18.5% (Wells Fargo Insurance Services, 2014).

Considering the proposed exponential growth of telemedicine as well as the inherent nature of EMS being distributed and relying on technology to facilitate care, telemedicine is rife with opportunity. Consequently, the purpose of this chapter is to elucidate telemedicine and how it is applied within EMS. With that being said, this chapter will proceed by first defining telemedicine, then describing the components and purposes of telemedicine, following with a discussion of the applications of telemedicine within EMS, and concluding with ideas for research.

WHAT IS TELEMEDICINE?

The term actually originates from the Greek word *tele*, which means at a distance, and the Latin word *mederi*, which is defined as healing. Thus, the literal translation means healing at a distance. The term *telemedicine*, however, was coined by Thomas Bird in the 1970s and has been defined as "the use of electronic information and communications technologies to provide and support health care when distance separates participants" (Field, 1996, p. 16). Similar to other complex concepts, telemedicine too has several definitions, such as "the practice of medicine without the usual physician–patient confrontation via an interactive audio-video communication system" (Bashshur et al., 2000, p. 614), "medical applications that use interactive video, typically for specialty or subspecialty physician consultants" (Field & Grigsby, 2002, p. 423), and "use of telecommunications technology for medical diagnostic monitoring and therapeutic purposes when distance and/or time separates the participants" (Hersh et al., 2006a, p. 3), to name a few.

Undoubtedly, these definitions have slight variations, yet it is evident that there are two common and defining characteristics of telemedicine. First, telemedicine inherently assumes distance or separation between two or more parties, and second, it utilizes telecommunication or information transmission technology (e.g., telephone, video, modem) to relay health information to foster clinical care (Bashshur et al., 2000). It should be noted, though, that participants no longer explicitly apply to physician and patient interactions as Bird initially suggested; that is, the participants refer to the original relationship of clinician and patient as well as multiple clinicians or even a combination of clinicians and a patient.

Due to the evolution and expansion of telemedicine, two types have emerged—asynchronous and synchronous (Smith, 2007). The first category, asynchronous applications, does not occur in real time but rather is known more commonly as "store and forward." Traditionally, information and medical data (e.g., images, audio, and text) in store-and-forward tools are captured, stored, and subsequently transmitted for later use (Hersh et al., 2006b). In fact, asynchronous applications do not rely on real-time discussion at all, but rather, they depend on interpretations and diagnosis to be conducted later. Additionally, store and forwards are predominantly personal computer based (Güler & Ubeyli, 2002); however, there are capabilities to be mobile and portable. Regardless if the technology is stationary or portable, the medical data that are transmitted at the receiving end is typically optimal quality such that it is generally equivalent to the sending end (Sable et al., 2009). Subsequently, they are extremely common in certain medical contexts, such as dermatology and radiology, with one review revealing that almost 50% of asynchronous studies were conducted within the field of dermatology (Deshpande et al., 2009). The popularization of such store-and-forward technology is attributed to evidence demonstrating improvements in patient management (Hersh et al., 2001), diagnostic accuracy, and patient satisfaction (Deshpande et al., 2009).

Clearly, research has provided evidence that asynchronous technologies have efficacious results in critical outcomes, which can possibly be connected to the advantages embedded or associated with such applications. For example, due to the nature of store-and-forward applications, they can function on a narrower bandwidth; therefore, they become easier to implement due to less complicated technological and systematic requirements (Perednia & Allen, 1995). Also, inherent in many asynchronous tools is the ability to annotate medical data (Allely, 1995), which too can be extremely informative and valuable in particular instances, especially for teleconsultations. To illustrate, a radiologist seeking another opinion can provide notations on a set of film for another colleague, which could possibly guide or alter the consultative process. Finally, another primary additive benefit of asynchronous uses is that they do not require the coordination of multiple individuals to be present at the same time (Hersh et al., 2006b). Consequently, asynchronous technologies afford for flexibility in the recipient's schedule by enabling the individual to examine the received patient data at his/her own convenience (Allely, 1995; Grigsby & Sanders, 1998). Given the workload and time commitments of clinicians as well as the reduction in hours for residents, the lack of person coordination is potentially a seriously underestimated asset.

Although store-and-forward techniques have positive characteristics, such as minimal people demands (i.e., they do not require multiple people at once), there are also detriments to utilizing these tools. For instance, asynchronous technologies demand better interface designs; interfaces must compensate for any deficits in design since store-and-forward tools do not have another human to interact with and address clinical requests in real time (Allely, 1995). Also, such technology is vulnerable to confidentiality issues; that is, store-and-forward applications are susceptible to breaches in security or improper patient disclosure (i.e., patient data sent to incorrect individuals; Grigsby & Sanders, 1998).

The second category, synchronous applications, includes real-time interactions between two or more parties. Synchronous technology is most notably used for videoconferencing, telementoring, and distance education, and evidence repeatedly seems reassuring. Systematic reviews suggest that these tools are associated with maintaining the same level of care compared to traditional face-to-face interventions (Barak et al., 2008; Currell et al., 2010) and even enhanced outcomes, such as diagnosis (Hersh et al., 2002), patient satisfaction, and reduced length of stay (Hersh et al., 2001).

Unquestionably, one of the primary advantages of synchronous applications is that they afford real-time, natural discussion and interactions (Loane et al., 2000), which enables requests to be addressed immediately and facilitate increased contact with other clinical experts. These real-time interactions are imperative for specific contexts that necessitate urgent yet accurate decision-making and procedural skills, such as videoconferencing in emergencies and surgical telementoring, which is characterized by a more prolific surgeon providing expertise to a colleague performing surgery at a separate location. Real-time interactions, embedded within synchronous technologies, are also useful for building rapport between physicians and patients and family members. The physician–patient relationship is integral considering that the interactions that comprise these relationships are negatively associated with "doctor shopping" (the act of finding an alternative caregiver; Kaplan et al., 1989) and positively related to medical information comprehension and recall (Ong et al., 1995), patient satisfaction, and compliance to treatment plans (Kaplan et al., 1989). Moreover, synchronous applications offer the capability of viewing live images (Smith et al., 2005), and for patient indicators such as heart rate or oxygen saturation levels, having a real-time assessment of such indicators can be pivotal for diagnosis and patient management. In addition, synchronous telemedicine provides more opportunities for education (Grigsby & Sanders, 1998) by fostering relationships between senior and junior colleagues, enabling regional locations to collaborate and discuss difficult patient cases, and providing individually focused learning (Sable et al., 2009).

On the other hand, synchronous telemedicine is not without its drawbacks. For instance, it relies on real-time interactions; thus, scheduling multiple individuals can be difficult and cumbersome. Moreover, real-time interactive feed eliminates the ability to obtain and save hard copies of medical data. Although saved medical data may not be necessary for all contexts, they can become a worthwhile referent, especially for complex cases. Also, such applications require greater technological and bandwidth requirements to be capable of producing a sufficient level of technical quality; in turn, visual images or auditory discussions may be reduced to merely acceptable levels (and potentially subpar) as opposed to optimal (Sable et al., 2009).

Both asynchronous and synchronous telemedicine can be used for a variety of purposes. Aiming to better understand and organize the expansive field of telemedicine, researchers have created taxonomies. Leveraging previous work, Tulu et al. (2005) proposed a taxonomy including two primary purposes of telemedicine: clinical and nonclinical. Expanding upon this taxonomy, Smith et al. (2005) further delineated the dominant purposes of telemedicine into three overarching categories: clinical, educational, and administrative. Within the clinical arena, telemedicine is used to supplement patient care, such as diagnostics and surgery. Meanwhile, for educational purposes, telemedicine is used to augment lectures, conferences, workshops, and grand rounds. Finally, for administrative applications, it is used to arrange interviews, meetings, and correspondence between regional facilities. See Table 10.1 for a summary of the advantages and disadvantages of asynchronous and synchronous telemedicine.

TABLE 10.1
Comparison of Asynchronous versus Synchronous Telemedicine

Type	Advantages	Disadvantages
Asynchronous	• Functions on narrower bandwidth • Mandates less complicated technological and systematic requirements • Easier to implement • Affords annotation • Offers flexibility in terms of scheduling users	• Demands better interface designs • Inability to address requests in real time • Vulnerable to confidentiality issues
Synchronous	• Affords real-time, natural discussion and interactions • Addresses feedback immediately • Builds rapport between users • Offers real-time assessment of fluctuating indicators (e.g., heart rate) and images • Facilitates education and collaboration between colleagues	• Demands coordinating schedules • Adds complexity to saving hard copies of images • Requires greater bandwidth

PURPOSES AND APPLICATIONS OF TELEMEDICINE IN EMS

Although we have primarily discussed telemedicine in general, it is not new to EMS settings. Despite being limited by technology capabilities, telemedicine has been utilized in some form or another for almost three decades, beginning with the transmission of EKGs via cellular phones in the late 1980s (Garza, 1998). We now live in a digital age defined by rapid technological advancements and the ubiquitous presence of devices needed to make modern work and life possible. Naturally, telemedicine is not exempt from the proliferation of new technologies (e.g., Google Glass; Carenzo et al., 2014), which have the capability of revolutionizing the field and making sci-fi-esque devices a reality. As a result, EMS telemedical systems vary widely in how they are designed and applied. Typical applications within this context range from mostly asynchronous, such as sending EKGs, still images, or video clips, to more synchronous, such as the sending and/or receiving of live feed audio and video (Bashford, 2011). While the potential applications of EMS telemedicine are as diverse as the technologies available, they may serve three common purposes: (1) documentation, (2) education, and (3) teleconsultation.

In principle, documentation is the most basic application of telemedicine within EMS as it does not require interaction between two or more providers. Documentation refers to materials that provide official information or evidence and serve as a record. There are a number of ways that EMS information can be documented using telemedicine. As already discussed, early applications of EMS telemedicine involved the transmission of EKGs from ambulance to hospital (Garza, 1998). This is one example of patient status in the prehospital setting. Another may be to take pictures or video of the patient upon arrival (Bergrath et al., 2013b) or record treatment refusals (Bashford, 2011).

The second purpose of EMS telemedicine is education. The information collected and transmitted using telemedicine can be used in the education and training of healthcare providers and emergency medical professionals (Bergrath et al., 2013a), either in real time or by using case studies in classroom-based instruction. For example, Haney et al. (2012) compared conventional lectures and tele-education lectures in managing wound care and found that both groups had similar knowledge acquisition and performance, indicating that tele-education is at least as effective as conventional education.

Finally, the arguably purest purpose of telemedicine is for consultation, coined *teleconsultation* or *telepresence*. Teleconsultation is potentially more inclusive of both asynchronous and synchronous technologies, whereas telepresence is restricted primarily to synchronous applications. Teleconsultation can potentially involve assessment and/or assistance. Assessment activities involve the evaluation of a patient prior to their arrival at the treatment facility. They include but are not restricted to initial diagnosis (Yperzeele et al., 2014), treatment decisions (Demaerschalk et al., 2010), and triage (i.e., assessment of the degree or urgency of wounds and illness; Brunetti et al., 2014). These activities have the potential benefit of cost reduction. For example, costs were reduced in a Spanish sample because fewer transport resources were used when telemedicine (i.e., transmission of EKGs and images) was utilized (Cabrera et al., 2002).

Teleconsultation may also be used to provide assistance to EMS providers in a variety of situations requiring varying levels of patient care and has been shown to be logistically feasible (Rortgen et al., 2013). Because EMS providers operate under a limited scope of practice, they are not always qualified to provide necessary care to patients. In more critical cases, teleconsultation may enable a remote physician to guide an EMS provider through life-saving treatment that an EMS provider would not be able to provide otherwise. Additionally, teleconsultation can allow for "treat and release" practices in which an EMS provider delivers the necessary care to a patient under the guidance of a remote physician (Widmer & Muller, 2014). Such practices may circumvent the unnecessary transport of patients to the emergency room (ER) (Haskins et al., 2002), thus reducing care costs and hospital staff workload.

We have highlighted some of the more common approaches to EMS telemedicine described in the published literature. However, there certainly exist more reasons and ways to implement telemedicine for EMS settings than those addressed here. EMS systems themselves can be diverse in nature in order to accommodate the laws and regulations set forth by the countries, states, and counties in which they operate and the demographics of the populations that they serve. As such, the needs and available resources of EMS systems will vary and consequently affect the adoption and implementation of telemedicine within each. Therefore, in the next section, we aim to describe some general guidance for researchers and practitioners considering EMS telemedicine applications.

GUIDANCE FOR RESEARCHERS AND PRACTITIONERS

When embarking on plans to implement or evaluate a telemedicine EMS system, there are a number of factors that both researchers and practitioners should consider in order to help prepare for the undertaking. These include (1) needs and objectives, (2) stakeholder buy-in, (3) resource availability, (4) privacy and confidentiality issues, and (5) use and maintenance policies. We next elaborate on each of these factors and offer a summary of tips for managing each in Table 10.2.

All decisions regarding EMS telemedicine should be related to a known need. Telemedicine implementation can be a complex and expensive process. The likelihood

TABLE 10.2

Success Factors and Tips for Implementing a Telemedicine EMS System

Success Factor	Implementation Tips
Needs and objectives	• Conduct a thorough need analysis prior to choosing or implementing a telemedicine EMS system • Match the telemedicine EMS system to the known need(s) to avoid resource waste and optimize success
Stakeholder buy-in	• Present compelling evidence (e.g., pilot testing, case studies of similar facilities) favoring the use of EMS telemedicine • Solicit stakeholder opinions about the telemedicine system • Address concerns and suspicions surrounding telemedicine prior to implementation • Include stakeholders in decision-making meetings and show that their contributions are valued
Resource availability	• Consider available resources (funds, time, labor, and space) prior to implementation • Identify ways in which EMS telemedicine might be a source of revenue, save resources, or prevent resource waste • Partner with another entity or department to pool resources • Utilize technologies that allow providers to perform normal duties while working with the telemedicine system to avoid needing additional personnel
Privacy and confidentiality	• Use products made to comply with HIPAA and other federal security and privacy requirements • Use advanced encryption to protect data • Store data on password-protected hard drives and servers • Keep telemedical equipment away from unauthorized visitors or staff
Use and maintenance	• Evaluate the telemedicine EMS system regularly and improve on weaknesses • Establish formal policies, procedures, and rules to establish proper-use practices • Share success stories associated with telemedicine • Build a cadre of staff champions to promote the use of EMS telemedicine • Make use fun: consider hosting contests to encourage buy-in and promote awareness of the telemedicine EMS system

of success is increased if plans are carefully constructed to make sure that the system was best able to meet user needs. As such, we recommend beginning with a requirement (i.e., needs) analysis, which will elucidate areas for improvement within the EMS system. Beginning with knowledge of what could be done better allows for more informed decisions about the type(s) of telemedicine that may be most appropriate and beneficial (Czaplik et al., 2014). Some telemedicine systems may sound really exciting, but if they do not meet a need within the organization or service area, then they are a waste of resources.

If it is decided that a certain telemedicine system is needed to address a specific gap, it will become necessary to obtain stakeholder buy-in. Stakeholders will include clinicians (from both prehospital and hospital settings) as well as organizational leadership (Bashford, 2011). Telemedicine can create suspicion in clinicians who might question how telemedicine will impact them personally. Clinicians who fear punitive action, for instance, might be resistant to the installation of an audio and video recording device onboard an ambulance. Although the purpose may be to document patient status prior to arrival in the ER, clinicians may have concerns that the video might be used to take punitive action against them. To encourage buy-in includes clinicians in the planning stages and addresses their concerns up front. Staff members are more likely to adopt new practices when they have been included in the planning and when their opinions have been considered.

In today's competitive environment, many organizations are trying to do more with less. It will be important to consider available resources in order to determine whether telemedicine is an appropriate and affordable solution. Buy-in from organizational leadership may be necessary to procure the needed resources (e.g., money) to implement telemedicine. It clearly communicates why telemedicine is needed and articulates the expected organizational benefits and explains other alternatives and what makes telemedicine the best option. Of course, resources are not restricted to funds. They also include the time, space, and workforce needed to support the telemedicine system. Remember to consider whether telemedicine can be implemented and sustained with the available workforce and space.

In healthcare, confidentiality and privacy issues are major concerns. Fortunately, Health Insurance Portability and Accountability Act (HIPAA) compliance can be achieved with advanced encryption to protect electronic information. However, encryption protects only the transmission of data. Additional consideration will be needed to ensure that data are protected once they have arrived. For instance, data will need to be kept in a space where they cannot be accessed by unauthorized visitors or staff (Bashford, 2011). Managing HIPAA compliance will depend on the type of telemedicine application adopted and the resources available, but both practitioners and researchers must consider these concerns and find workable solutions.

Finally, adoption of telemedicine will require policies regarding how the system is to be used, tested, and maintained (Bashford, 2011). Once again, these policies and procedures will vary in accordance with the type of telemedicine application selected. They may evolve with use and trial and error, but researchers and practitioners should make initial plans for use and maintenance prior to implementation of telemedicine. These plans should include details for ongoing assessment and evaluation of a system's efficacy and utility.

Much of the research surrounding telemedicine in EMS is dedicated to determining the feasibility of implementing it within a given context, hospital, or EMS system (e.g., Wu et al., 2014). Although such studies are indeed important as an initial first step in determining whether telemedicine can be used to solve a known or perceived deficit in patient care, it is also necessary to conduct research with the purpose of elucidating the benefits that telemedicine in EMS holds for patients, providers, organizations, and communities in terms of care quality, safety, and costs. These types of evaluative studies can inform development and implementation of telemedicine in EMS and help to determine their value to the system. Initial evidence is positive but far from conclusive about the positive implications of telemedicine. Moreover, because there are so many different combinations of applications and purposes of EMS telemedicine, it is especially important to share results, findings, and lessons learned from adoption of telemedicine. Doing so allows others for triangulation of information.

CONCLUSIONS

EMS is characterized by distributed providers who must seamlessly integrate and coordinate to deliver quality patient care. Telemedicine is one mechanism to facilitate the synthesis of this distributed care. Realizing the potential of telemedicine, EMS has utilized such technology primarily for three purposes: documentation, education, and teleconsultation. As established, there is some evidence to demonstrate the promise of telemedicine within EMS (e.g., Bolle et al., 2009); however, most studies are dedicated toward establishing feasibility (Bergrath et al., 2013a; Yperzeele et al., 2014). Given that telemedicine is gaining acceptance within EMS and the production of innovative technology is rapid, the possibilities will broaden mandating the need for future research. For example, how does technology change the provision of patient care? What advantages or disadvantages do patients experience from specific telemedical hardware and software? How is taskwork and teamwork altered with the addition of telemedicine? How do multiteam systems adapt to the presence of telemedicine? Does telemedicine enable organizations to reduce resource waste and cost by better equipping clinicians? What detriments or benefits can communities experience with the addition of telemedicine?

Although these examples simply scratch the surface of the unanswered questions, we encourage researchers and practitioners to collaborate to better navigate these unexplored domains. These questions and others are undoubtedly difficult and complex; however, patients deserve answers to these questions. Evidence should drive change in care; change should not be accelerated exclusively by flashy technology.

REFERENCES

Allely, E.B. (1995). Synchronous and asynchronous telemedicine. *Journal of Medical Systems*, *19*, 207–212.
Barak, A., Hen, L., Boniel-Nissim, M. & Shapira, N. (2008). A comprehensive review and a meta-analysis of the effectiveness of Internet-based psychotherapeutic interventions. *Journal of Technology in Human Services, 26*, 109–160.

Bashford, C. (2011). *Thinking about EMS telemedicine?* Retrieved from http://www.general -devices.com/files/learning_pdf/EMS_World_Article.pdf.

Bashshur, R.L., Reardon, T.G. & Shannon, G.W. (2000). Telemedicine: A new health care delivery system. *Annual Review of Public Health, 21,* 613–637.

Bergrath, S., Czaplik, M., Rossiant, R., Hirsch, F., Beckers, S.F., Valentin, B. et al. (2013a). Implementation phase of a multicenter prehospital telemedicine system to support paramedics: Feasibility and possible limitations. *Scandinavian Journal of Trauma, Resuscitation, and Emergency Medicine, 21,* 1–10.

Bergrath, S., Rossaint, R., Lenssen, N., Fitzner, C. & Skorning, M. (2013b). Prehospital digital photography and automated image transmission in an emergency medical service: An ancillary retrospective analysis of a prospective controlled trial. *Scandinavian Journal of Trauma, Resuscitation & Emergency Medicine, 21,* 1–9.

Bolle, S.R., Larsen, F., Hagen, O. & Gilbert, M. (2009). Video conferencing versus telephone calls for team work across hospitals: A qualitative study on simulated on simulated emergencies. *Biomed Central Emergency Medicine, 9,* 22.

Brunetti, N.D., Dellegrottaglie, G., Lopriore, C., Di Giuseppe, G., de Gennaro, L., Lanzone, S. et al. (2014). Prehospital telemedicine electrocardiogram triage for a regional public emergency medical service: Is it worth it? A preliminary cost analysis. *Clinical Cardiology, 37,* 140–145.

Cabrera, M.F., Arredondo, M.T. & Quiroga, J. (2002). Integration of telemedicine into emergency medical services. *Journal of Telemedicine and Telecare, 8,* 12–14.

Carenzo, L., Barra, F.L., Ingrassia, L., Colombo, D., Costa, Al. & Corte, F.D. (2014). Disaster medicine through Google Glass. *European Journal of Emergency Medicine, 22,* 222–225.

Currell, R., Urquhart, C., Wainwright, P. & Lewis, R. (2010). Telemedicine versus face to face patient care: Effects on professional practice and health care outcomes. *Cochrane Database of Systematic Reviews, 2,* 1–35.

Czaplik, M., Bergrath, S., Rossaint, R., Thelen, S. Brodziak, T., Valentin, B. et al. (2014). Employment of telemedicine in emergency medicine. *Methods of Information Medicine, 53,* 99–107.

Demaerschalk, B.M., Bobrow, B.J., Raman, R., Kiernan, T.J., Aguilar, M.I., Ingall, T.J. et al. (2010). Stroke team remote evaluation using a digital observation camera in Arizona: The initial Mayo Clinic experience trial. *Stroke, 41,* 1251–1258.

Deshpande, A., Khoja, S., Lorca, J., Mckibbon, A., Rizo, C., Husereau, D. et al. (2009). Asynchronous telehealth: A scoping review of analytic studies. *Open Medicine, 3,* 69–91.

Field, M.J. (1996). *Telemedicine: A guide to assessing telecommunications in health care.* Washington, DC: National Academies Press.

Field, M.J. & Grigsby, J. (2002). Telemedicine and remote patient monitoring. *Journal of American Medical Association, 288,* 423–425.

Garza, M.A. (1998). Telemedicine: The key to expanded EMS or an expensive experiment? *JEMS, 23,* 28–38.

Grigsby, J. & Sanders, J.H. (1998). Telemedicine: Where it is and where it's going. *Annals of Internal Medicine, 129,* 123–127.

Güler, N.F. & Ubeyli, E.D. (2002). Theory and applications of telemedicine. *Journal of Medical Systems, 26,* 199–220.

Haney, M., Silvestri, S., Van Dillen, C., Ralls, G., Cohen, E. & Papa, L. (2012). A comparison of tele-education versus conventional lectures in wound care knowledge and skill acquisition. *Journal of Telemedicine and Telecare, 18,* 79–81.

Haskins, P.A., Ellis, D.G. & Mayrose, J. (2002). Predicted utilization of emergency medical services telemedicine in decreasing ambulance transports. *Prehospital Emergency Care, 6,* 445–448.

Hersh, W.R., Helfand, M., Wallace, J.A., Kraemer, D., Patterson, P., Shapiro, S. et al. (2002). A systematic review of the efficacy of telemedicine for making diagnostic and management decisions. *Journal of Telemedicine and Telecare, 8,* 197–209.

Hersh, W.R., Hickam, D.H., Severance, S.M., Dana, T.L., Krages, K.P. & Helfand, M. (2006a). Diagnosis, access and outcomes: Update of a systematic review of telemedicine services. *Journal of Telemedicine and Telecare, 12,* 3–31.

Hersh, W.R., Hickam, D.H., Severance, S.M., Dana, T.L., Krages, K.P. & Helfand, M. (2006b). *Telemedicine for the Medicare population: Update.* Rockville, MD: Agency for Healthcare Research and Quality.

Hersh, W.R., Wallace, J.A., Patterson, P.K., Kraemer, D.F., Nichol, W.P., Greenlick, M.R. et al. (2001). *Telemedicine for the Medicare population: Pediatric, obstetric, and clinician-indirect home interventions in telemedicine.* Rockville, MD: Agency for Healthcare Research and Quality.

Kaplan, S.H., Greenfield, S. & Ware, J.E. (1989). Assessing the effects of physician–patient interactions on the outcomes of chronic disease. *Medical Care, 27,* S110–S127.

Loane, M.A., Bloomer, S.E., Corbett, R., Eedy, D.J., Hicks, N., Lotery, H.E. et al. (2000). A comparison of real-time and store-and-forward teledermatology: A cost–benefit study. *British Journal of Dermatology, 143,* 1241–1247.

Ong, L.M.L., De Haes, J.C.J.M., Hoos, A.M. & Lammes, F.B. (1995). Doctor–patient communication: A review of the literature. *Social Science & Medicine, 40,* 903–918.

Perednia, D.A. & Allen, A. (1995). Telemedicine technology and clinical applications. *Journal of American Medical Association, 273,* 483–488.

Roashan, R. (2014). *A dedicated study on telehealth that provides detailed analysis of the world market.* Englewood, CO: IHS Pressroom.

Rortgen, D., Bergrath, S., Rossaint, R., Beckers, S.K., Fischermann, H., Na, I. et al. (2013). Comparison of physician staffed emergency teams with paramedic teams assisted by telemedicine: A randomized, controlled simulation study. *Resuscitation, 84,* 85–92.

Sable, C., Reyna, M. & Holbrook, P.R. (2009). Telemedicine applications in pediatrics. In C.U. Lehmann, G.R. Kim & K.B. Johnson (Eds.), *Pediatric informatics: Computer applications in child health* (pp. 279–292). New York, Springer.

Smith, A.C. (2007). Telemedicine: Challenges and opportunities. *Expert Review of Medical Devices, 4,* 5–7.

Smith, A.C., Bensick, M., Armfield, N., Stillman, J. & Caffery, L. (2005). Telemedicine and rural health care applications. *Journal of Postgraduate Medicine, 51,* 286–293.

Tulu, B., Chatterjee, S. & Laxminarayan, S. (2005). A taxonomy of telemedicine efforts with respect to applications, infrastructure, delivery tools, type of setting and purpose. *Proceedings of the 38th Hawaii International Conference on System Science,* 1–10.

Wells Fargo Insurance Services (2014). *The growing trend toward telehealth.* Retrieved from https://wfis.wellsfargo.com/insights/whitepapers/Documents/WP_TheGrowingTrend TowardTelehealth.pdf.

Widmer, A. & Mimuller, H. (2014). Using Google Glass to enhance pre-hospital care. *Swiss Medical Informatics, 30.* Retrieved from http://www.medical-informatics.ch/index.php /smiojs/article/viewFile/316/359.

Wu, T.C., Nguyen, C., Ankrom, C., Yang, J., Persee, D., Vahdiy, F. et al. (2014). Prehospital utility of rapid stroke evaluation using in-ambulance telemedicine: A pilot feasibility study. *Stroke, 45,* 2342–3247.

Yperzeele, L., Van Hooff, R., De Smedt, A., Espinoza, A.V., Van Dyck, R., Van de Casseye, R. et al. (2014). Feasibility of ambulance-based telemedicine (FACT) study: Safety, feasibility and reliability of third generation in-ambulance telemedicine. *PLoS ONE, 9,* e110043.

11 Communication and Patient Care Handover
Prehospital Emergency Preparation

Heather C. Lum and Shane E. Halse

CONTENTS

INTRODUCTION

EMS professionals, who are medically trained first responder personnel, act as a collaborative unit in their effort to care for the sick and injured patients being transported from the scene of an incident to the hospital. In this short window of time, there is a plethora of information that the EMS staff are collecting about the patient's history, vitals, and changing medical state, all of which have to be documented and handed over to the ED staff upon arrival at the hospital. In an ideal world, this handover process is smooth, succinct, and thorough. However, this is not always the case. In this chapter, the patient handoff process will be examined, along with the common pitfalls and areas in which improvements are needed.

In an emergency medical situation, EMS professionals and first responders, including firefighters, paramedics, and EMTs, are the primary individuals on the scene who will assess and care for the patient. In a time-critical and stressful environment, these individuals are responsible for gathering as much information as possible about the patient. This is particularly difficult due to how diverse the patients are and how unique the situation and circumstances may be. Patients reflect all sections of society and can be of any age, race, or ethnicity, with a plethora of different acute or chronic illnesses or injuries to account for. Further, the scene of an emergency can take place almost anywhere—be it a burning building, the floor of an office, or the side of a road. Therefore, information collection and processing occur from the moment that the EMS staff arrives on the scene and must be quick and accurate. This information is collected and broken down into different criteria and includes the types shown in Table 11.1.

This information not only is often recorded in written form on patient care sheets but can also be verbally shared with the ED staff upon arrival at the hospital. The handover process between EMS and ED is often performed quickly due to the time-sensitive nature of getting the patient to the next stage of their care. Figure 11.1 is a summary of how the handover process works from beginning to end.

TABLE 11.1
Prehospital EMS Checklist

Type of Information	Examples
Mechanism of injury/illness	Respiratory distress
	Ejection from a vehicle
	Penetrating injury to the trunk
	Gunshot wounds
	Burns
	Punctures
Medical/vitals	Airway compromise
	Severe hemorrhage
	Cardiac chest pain
	Anaphylaxis
	Unconsciousness
	Cardiogenic shock
	Blood pressure
Medical history	Previous/similar episodes
	Medications
	Allergies
	General demographics such as age and gender
Treatments given	Any medications given
	Any procedures started

Source: Adapted from Welsh Ambulance Services NHS Trust, *WAST Standard Operating Procedure: Hospital Pre-alerting*, Welsh Ambulance Services NHS Trust, Denbighshire, UK, 2010.

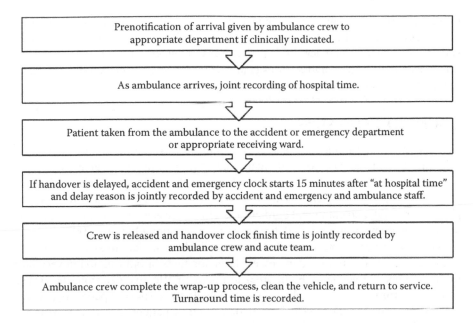

FIGURE 11.1   Flow chart of handover process. (Adapted from Gatling, E., and Ansell, J., *Ensuring Timely Handover of Patient Care—Ambulance to Hospital*, National Health Service, London, 2008.)

In an ideal scenario, this handover process is performed smoothly with all of the pertinent and necessary details about the patient being shared between the EMS providers and the ED. However, given what we know about human cognition and teamwork, the patient may be at the receiving end of the adverse effects of communication and information transfer failures. This breakdown in communication can happen from multiple fronts and for a variety of reasons. In this chapter, this handover process will be examined including the challenges in communication, the pitfalls of the process, and best practices for improved handover process.

TIME PRESSURE

Eckstein and Chan (2004) examined the effect of emergency room (ER) crowding on ambulance availability and off-load time. Astonishingly, in Los Angeles, they found that one out of every eight transports yielded a transfer wait time of at least 15 minutes, with 8.4% of incidents yielding a wait time greater than 1 hour. What this means is that ambulances and their EMS staff are effectively grounded and out of service until they can complete the handover process. This results in a loss of nearly five million hours of EMS system productivity in a given year (Williams, 2006). Also, this impacts the patient's quality of care including delay of definitive care, poor pain control, delayed time to antibiotics, and prolonged hospital stay. To combat this problem, several states have attempted to create a more streamlined and efficient handover process. In some instances, patient transfer time has been reduced to only a few minutes. While this means a reduced wait time for patients, it also leads to increased time pressure for EMS staff to transfer the vital patient information to the hospital staff in an effective manner. Information transfer under time pressure and stress often leads to information loss (Pries et al., 2003). This is expanded upon further in the following sections regarding cognition and stress.

COGNITION AND STRESS

The relationship between stress and cognitive processes is not a new one. However, it is only within the past 30 to 40 years that more definitive explanations for this phenomenon have surfaced (Sandi, 2013). Stress is the perception that situations are physically and psychologically dangerous (Atkinson et al., 1996). It is the result of physical, mental, and emotional burdens; therefore, stress is impacted by an individual's explanation, evaluation, and direction of the situation (Altunas, 2003). Because stress is so pervasive and influential, decision-making under stress is a vital tool, because individuals are forced to process large amounts of information in very little time and must act accordingly. In order to make an effective decision, the correct context and timing must be applied appropriately. In the context of patient handovers, the task of transferring critical information about a patient in a timely matter is inherently a stressful one (Solet et al., 2005).

Under a significant amount of stress, our decision-making abilities can be negatively impacted and we are likely to make incorrect decisions possibly leading to mistakes. When people experience a great deal of stress, they tend to possess tunnel vision, which leads to poor decision-making (Dirkin, 1983). Much of the research on stress conducted in the laboratory typically finds that stress hinders decision-making when individuals become overloaded with information (Harris et al., 2005). Stress or arousal has been a defining factor that has been studied through its effects on cognition and performance. For nearly a century, this relationship has been explained through the use of the Yerkes–Dodson law. This law is described graphically as an inverted U shape with the optimal performance coinciding with a medium amount of arousal. Performance is diminished when the arousal level is too low or too high (Yerkes & Dodson, 1908). This view has been criticized by researchers such as Michael Mendl (1999) for being too simplistic to explain the complex relationship between stress and cognition. Mendl does concede, however, that this law may be a good heuristic for unifying a theory of stress and cognition. For instance, Preston et al. (2007) found that participants who experienced stress took longer to develop advantageous decision-making strategies. What this means for the handover process is that stress can affect the ability to properly transfer critical and time-sensitive information between the EMS professionals and the ER staff. The process and considerations of information transfer are described in further detail in the next section.

COMMUNICATION AND INFORMATION TRANSFER

Information transfer, regardless of the context, is an exercise in communication between two or more entities. In order to ensure adequate transmission of information, it must be encoded and stored, accurately remembered, and accurately communicated. Additionally, the team and communication literature has established a number of factors that influence the accuracy and reliability of information transfer during time-sensitive scenarios. In the following section, age of information, attentional focus, distractors, explicit communication, and consistency are examined as they relate to the handover process.

AGE OF INFORMATION

Several cognitivists, including Baddeley (1976), have expounded on the notion of the information age or the idea that information is subject to forgetting. The information age can be thought of as a shift from knowing data offhand to not knowing offhand but rather being able to search for it through technological means. The information age is largely dependent on transfer from short-term to long-term memory, rehearsal, and elapsed time. Likelihood of error has been shown to increase with the age of information (Salthouse, 1996).

ATTENTIONAL FOCUS

We can only attend to so much in the world. One way to increase the efficacy of attentional focus is through the use of team-based practices. This involves all members of the response team focusing their attention on information as it is generated, rather than only one person doing so. Salas et al. (2005) found that this common attentional focus leads to backup behavior among the group, which ultimately leads to increased accuracy in recall. Backup behavior has generally been defined as helping other team members perform their roles under high levels of workload and is thought to be critical for effective team performance (Porter et al., 2003). In this instance, backing up other's behavior ensures that information is indeed transferred through multiple sources. However, ED personnel are often attending to multiple tasks while the EMS staff is giving the verbal handover (Owen et al., 2009). This lack of active listening can severely compromise the ability to encode and retain this information later. By promoting backup behavior, limitations in attentional focus can be mediated and overcome.

DISTRACTORS

One assumption made in real-world scenarios is that there is always noise in the system; that is, no system is devoid of distractions and it is these distractions that make up the noise. Such noise diverts attention away from the information that is being generated and transferred, which can lead to problems in encoding and storing such information. This noise can be from relevant sources (e.g., another piece of vital information being said at the same time) or irrelevant noise (e.g., alarms going off, others talking in the background). In any effect, there is a plethora of research that has found that noise impairs the recall of information even if the particular noises are well known (Rabbitt, 1968; Murthy et al., 1995). While no system can be completely noise-free, the idea of reducing the noise can improve the recall of the information. Training responders to ignore distractions, actively reduce noise, or filter distractions out can result in reduced overall noise.

EXPLICIT COMMUNICATION AND EASE OF INFORMATION TRANSFER

If transfer is performed through a direct retrieval method, it should yield better encoding and retrieval. That is, the more that information is presented in a clear, concise, and direct manner, the easier it will be to understand it. This is called the transfer-appropriate processing theory, in which information retrieval is facilitated if the encoding corresponds to the way that the information is to be retrieved (Brown, 1997). Also, there is often a lack of common language between the EMS staff and the hospital staff. It was reported that the ER nurses, when unable to decipher the verbal handover clearly due to lack of common language, instead used other means such as their own patient assessment and written reports (Owen et al., 2009). Even when this information is in written form, it may be omitted or altered during transfer if nurses are unable to read that information or they choose to ignore it. In one study, researchers investigated this discrepancy and found that of 100 records, 26 had some form of discrepancy. These discrepancies included previous medical history, frequency of event, timing of events, allergies, drugs taken, and others (Murray et al., 2012). By empirically investigating these discrepancies, we should be able to resolve them by generating a better understanding between the involved parties. For example, if terminology and lingo are an issue, then establishing a standard for jargon could aid in guiding the words used during transfer. This solution would also help alleviate the issue of consistency in which many different words are used each with similar meanings or in which one word may have multiple meanings.

CONSISTENCY

The other issue with information transfer is the lack of consistency across county, state, and country. In other words, regulations vary as to what information is written and can be reported via radio, and what can be verbally shared is often quite different between units. Again, this can lead to a lack of common language and be frustrating for the individuals trying to transfer the information as well as the receiver of said information. A set of nationwide or worldwide standard practices could help systematize performance and maintain consistency across units.

IMPROVING THE HANDOVER PROCESS

Clearly, in many handover cases, there are deficiencies and adverse effects that occur when the process is not performed optimally. Although a critical component in patient care, there are gaps in the handover process that are in need of attention and improvement. There has been a plethora of research on this process but only a limited number that have focused on possible solutions. From the research that is available, it is clear that special attention should be made to the communication of necessary information between EMS and the ED staff. More specifically, recommendations for improvements to the handover process include guideline development to standardize the process, greater use of information technology facilities, ongoing feedback to staff, and quality assurance and educational training activities.

TIME–ACCURACY TRADEOFF

There is a tricky balance between having a brief and thorough handover. As Hayden et al. (2001) discussed, if a handover is too brief, then important information may not be passed. This can lead to a "negative handover," in which essential information is not documented and that information is then lost (Tobin et al., 2000). What is documented needs to be accessible and legible and to conform to the standardization process for acronyms and protocol. Standardized protocols ensure that vital information is not missed and also streamline the process to aid in faster handover times while allowing for the transition of more information. Indeed, researchers have found that using handover protocols reduces the time spent conducting the handoffs while also increasing the amount of information being transferred (Lazzara et al., 2014).

ELECTRONIC REPORTS

One possible means to facilitate quicker information transfer may be the use of electronic patient reports. Raptis et al. (2009) compared paper-based and electronic-based medical reports and found that those who used electronic records achieved better continuity and higher amount and accuracy of information. However, this study was conducted for transfer of care information between hospital night teams. In general, many of the studies that have looked at electronic reports focus on those patients already admitted (Mandl et al., 1998). What is lacking is research focused on the more time-critical nature of information transfer and reports in the handover process between EMS and ED staff. As Talbot and Bleetman (2007) mention, electronic patient report forms are currently under development and may provide a partial solution for the transfer of prehospital information to ED staff. The key here is that electronic records may be a partial solution. While technology may help, it is still essential to focus on communication and protocol training so that important prehospital information is not lost or misinterpreted regardless of the record keeping method.

USE OF CHECKLISTS AND STANDARDIZATION

Where electronic patient reports may not be a viable solution yet, checklists are a method for improving team performance when routine tasks can be identified. For example, a checklist for verifying the safety of inserting intravenous lines lowered the infection rate by 11% at one institution, preventing an estimated eight deaths and saving $2 million in just over 1 year (Wu et al., 2002). These checklists could be designed in a traditional format or electronic format depending on the needs and resources of the adopting agency. A redesigned electronic checklist should allow EMS crews at the accident scene and, during transport, to rapidly enter information that is essential for the ED team. The benefit of a checklist is that it reminds the EMS crew of what is needed and limits their effort to collecting essential information only.

Use of standardized protocols is one clear recommendation that seemed to resonate throughout the research literature on transitions of care (Riesenberg et al., 2009). Standardization aids providers in regard to what information is reported and how reports are prepared. It is especially important to make communication events as concise as possible to limit the time spent on erroneous or unnecessary information being collected and shared (Penner et al., 2003). Additionally, it may be prudent for multiple or adjoining counties or jurisdictions to adopt the same protocol. This can be accomplished through a reporting system such as E-STAT, which stands for events prior (why EMS was called), subjective findings, triage/time, allergies/assessment, and treatment (Thakore & Morrison, 2001). EMS staff often service multiple hospitals and areas; therefore, it is essential that they are able to "speak the same language" to any ED about the essential patient information regardless of the location.

TRAINING AND TEAM TRAINING

The importance of proper training cannot be understated. Training has been shown to combat issues with improper or inaccurate handover and communication discrepancies. There is currently a lack of formal education in patient handover at all levels of training (Manser & Foster, 2011). Through standardization and improvement in information records, the handover process can be improved. But these efforts must work in concert with training of both the EMS and the ED staff. Communication training, clinical team leadership, and team discipline must support the communication process between EMS crews and the ED team to ensure that important prehospital information is not lost or misinterpreted.

CONCLUSION

The handover process is a vital piece of the puzzle when it comes to patient care due to the multiple deficiencies and issues that can occur in communication between EMS and hospital staff. For instance, one study revealed that in Australian emergency departments, nearly 10% of patients may be adversely affected due to the result of poor communication (Ye et al., 2007). In the United States, preventable medical errors are abundant, and 70% of those are due to communication breakdowns (Joint Commission, 2012). Likewise, timely and effective communication between EMS and the receiving hospital has the potential to save lives and improve patient eligibility for time-sensitive therapies such as stroke medications and other thrombolytic therapies (Abdullah et al., 2008). What is needed is a systematic investigation of the problems reported in the handover process and the solutions that will yield the more fruitful results. While this has been conducted in other aspects of patient handover such as nursing and physician handoffs within the hospital (Riesenberg et al., 2009, 2010), further investigation is needed in the prehospital context given the unique nature of the task.

Human factors science is integral to interdisciplinary research aimed at understanding and improving patient handover (Harvey et al., 2007). The contribution of human factors science in supporting current research and improvement efforts ranges from the definition and measurement of technical as well as nontechnical skills related to patient handovers, to the design, implementation, and evaluation of handover processes and supporting tools as well as targeted educational interventions. Effective communication and teamwork are essential for the provision of high-quality and safe patient care. Of course, the aim of both EMS professionals and the ED staff is always smooth and efficient patient transfer. Placing an emphasis on training with structured communication tools can decrease the risk of communication failures. Understanding the importance of effective communication and what it contributes to the continuity of patient care and enhanced patient safety is indispensable for the EMS provider. A summary of these key points is found in Table 11.2.

TABLE 11.2
Summary of Key Points

Issue	Description
Communication and Information Transfer	
Age of information	One technique to avoid this is to have critical information presented closer to the time of patient transfer, thus reducing the time elapsed. In addition, critical information should also be readily available and require updates regularly, which would assist in the rehearsal and allowing responders to move these data from short-term to long-term memory.
Attentional focus	A technique to reduce attentional focus could be to use three-way communication.
Distractors	While no system can be completely noise-free, the idea of reducing the noise can improve the recall of the information. Training responders to ignore the distractions and noise or, rather, to filter the distractions out can result in reduced overall noise.
Explicit communication and ease of information transfer	By investigating these discrepancies, we would be able to resolve them by generating a better understanding between the involved parties.
Consistency	Consistency in terminology and practice would also help alleviate the issue of consistency in which many different words are used each with similar meanings or in which one word may have multiple meanings.
Improving the Handover Process	
Standardization	Develop or adopt a common lingo in which EMS and EDs can communicate and thus speak the same language.
Time–accuracy tradeoff	Standardized protocols ensure that vital information is not missed and also streamline the process to aid in faster handover times while allowing for the transition of more information.
Electronic reports	This is only a partial solution, and it is still essential that to focus on communication and protocol training be the focus so that important prehospital information is not lost of or misinterpreted regardless of the record keeping method.
Checklists	The benefit of a checklist is that it reminds the EMS crew of what is needed and limits their effort to collecting essential information only.
Training and team training	Communication training, clinical team leadership, and team discipline must support the communication process between EMS crews and the ED team to ensure that important prehospital information is not lost or misinterpreted.

REFERENCES

Abdullah, A.R., Smith, E.E., Biddinger, P.D., Kalenderian, D. & Schwamm, L.H. (2008). Advance hospital notification by EMS in acute stroke is associated with shorter door-to-computed tomography time and increased likelihood of administration of tissue-plasminogen activator. *Prehospital Emergency Care, 12*, 426–431.

Atkinson, R.L., Atkinson, R.C., Smith, E.E., Bem, D.J. & Nolen-Hoeksema, S. (1996). *Introduction to psychology*. Ankara, Turkey: Arkadas Publishing.

Altunas, E. (2003). *Stress management*. Istanbul, Turkey: Alfa Publishing.

Baddeley, A.D. (1976). *The psychology of memory*. New York: Basic Books.

Brown, N.R. (1997). Context memory and the selection of frequency estimation strategies. *Journal of Experimental Psychology: Learning, Memory, and Cognition, 23*, 898–914.

Dirkin, G.R. (1983). Cognitive tunneling: Use of visual information under stress. *Perceptual and Motor Skills, 56*, 191–198.

Eckstein, M. & Chan, L.S. (2004). The effect of emergency department crowding on paramedic ambulance availability. *Annals of Emergency Medicine, 43*, 100–105.

Gatling E. & Ansell, J. (2008, October). *Ensuring timely handover of patient care—Ambulance to hospital*. National Health Service, London.

Harris, W.C., Hancock, P.A. & Harris, S.C. (2005). Information processing changes following extended stress. *Military Psychology, 17*, 115–128.

Harvey, C.M., Schuster, R.J., Durso, F.T., Matthews, A.L. & Surabattula, D. (2007). Human factors of transition of care. In P. Carayon (Ed.), *Handbook of human factors and ergonomics in health care and patient safety* (pp. 233–248). Mahwah, NJ: Lawrence Erlbaum Associates.

Hayden, S., Vilke, G., Thierbach, A. & Surgue, M. (2001). In E. Soriede & C. Grande (Eds.) (2001). *Pre-hospital trauma care*. New York: Marcel Dekker.

Joint Commission. (2012). *Sentinel event statistics data: Root causes by event type*. http://www.jointcommission.org/assets/1/18/Root_Causes_by_Event_Type_2004-2Q2013.pdf. Accessed June 30, 2014.

Lazzara, E.H., Riss, R., Chan, Y.R., Keebler, J.R., Palmer, E., Smith, D.S. et al. (2014, October). Comparing the effectiveness of handoff tools. In K. Fletcher (Chair), *Medical team handoffs: Current and future directions*. Panel conducted at the 58th Annual Meeting of the Human Factors and Ergonomics Society, Chicago.

Mandl, K.D., Kohane, I.S. & Brandt, A.M. (1998). Electronic patient–physician communication: Problems and promise. *Annals of Internal Medicine, 129*, 495–500.

Manser, T. & Foster, S. (2011). Effective handover communication: An overview of research and improvement efforts. *Best Practice & Research Clinical Anaesthesiology, 25*, 181–191.

Mendl, M. (1999). Performing under pressure: Stress and cognition. *Applied Animal Behavior Science, 65*, 221–244.

Murray, S.L., Crouch, R. & Ainsworth-Smith, M. (2012). Quality of the handover of patient care: A comparison of pre-hospital and emergency department notes. *International Emergency Nursing, 20*, 24–27.

Murthy, V.S., Malhotra, S.K., Bala, I. & Raghunathan, M. (1995). Detrimental effects of noise on anesthetists. *Canadian Journal of Anesthesia, 42*, 608–611.

Owen, C., Hemmings, L. & Brown, T. (2009). Lost in translation: Maximizing handover effectiveness between paramedics and receiving staff in the emergency department. *Emergency Medicine Australia, 21*, 102–107.

Penner, M.S., Cone, D.C. & MacMillan, D. (2003). A time-motion study of ambulance-to-emergency department radio communications. *Prehospital Emergency Care, 7*, 204–208.

Porter, C.O., Hollenbeck, J.R., Ilgen, D.R., Ellis, A.P., West, B.J. & Moon, H. (2003). Backing up behaviors in teams: The role of personality and legitimacy of need. *Journal of Applied Psychology, 88*, 391.

Preston, S.D., Buchanan, T.W., Stansfield, R.B. & Bechara, A. (2007). Effects of anticipatory stress on decision making in a gambling task. *Behavioral Neuroscience, 121*, 257–263.

Pries, A.R., Reglin, B. & Secomb, T.W. (2003). Structural response of microcirculatory networks to changes in demand: Information transfer by shear stress. *American Journal of Physiology-Heart and Circulatory Physiology, 284*, H2204–H2212.

Rabbitt, P.M. (1968). Channel-capacity, intelligibility and immediate memory. *Quarterly Journal of Experimental Psychology, 21*, 241–248.

Raptis, D.A., Fernandes, C., Chua, W. & Boulos, P.B. (2009). Electronic software significantly improves quality of handover in a London teaching hospital. *Health Informatics Journal, 15*, 191–198.

Riesenberg, L.A., Leitzsch, J. & Cunningham, J.M. (2010). Nursing handoffs: A systematic review of the literature. *American Journal of Nursing, 110*, 24–34.

Riesenberg, L.A., Leitzsch, J. & Little, B.W. (2009). Systematic review of handoff mnemonics literature. *American Journal of Medical Quality, 24*, 196–204.

Salas, E., Sims, D.E. & Burke, C.S. (2005). Is there a "big five" in teamwork? *Small Group Research, 36*, 555–599.

Salthouse, T.A. (1996). The processing-speed theory of adult age differences in cognition. *Psychological Review, 103*, 403.

Sandi, C. (2013). Stress and cognition. *Wiley Interdisciplinary Reviews: Cognitive Science, 4*, 245–261.

Solet, D.J., Norvell, J.M., Rutan, G.H. & Frankel, R.M. (2005). Lost in translation: Challenges and opportunities in physician-to-physician communication during patient handoffs. *Academic Medicine, 80*, 1094–1099.

Talbot, R. & Bleetman, A. (2007). Retention of information by emergency department staff at ambulance handover: Do standardized approaches work? *Emergency Medical Journal, 24*, 539–542.

Thakore, S. & Morrison, W. (2001). A survey of the perceived quality of patient handover by ambulance staff in the resuscitation room. *Emergency Medicine Journal, 18*, 293–296.

Tobin, M., Nguyen-Van-Tam, J., Bailey, R., Pearson, J., Dove, A. & Durston, P. (2000). Completeness of recording clinical information and diagnostic accuracy of ambulance crews using scanned patient report forms. *Prehospital Immediate Care, 4*, 143–147.

Welsh Ambulance Services NHS Trust. (2010). *WAST standard operating procedure: Hospital pre-alerting*. Welsh Ambulance Services NHS Trust, Denbighshire, UK, 2010.

Williams, D.M. (2006). 2005 JEMS 200 city survey. *Journal of Emergency Medical Services, 31*, 44–100.

Wu, A., Pronovost, P. & Morlock, L. (2002). ICU incident reporting systems. *Journal of Critical Care, 17*, 86–94.

Ye, K., Taylor, D., Knott, J.C., Dent, A. & MacBean, C.E. (2007). Handover in the emergency department: Deficiencies and adverse effects. *Emergency Medicine Australasia, 19*, 433–441.

Yerkes, R.M. & Dodson, J.D. (1908). The relation of strength of stimulus to rapidity of habit formation. *Journal of Comparative Neurology and Psychology, 18*, 459–482.

12 Changes from Within

How Paramedic Services Can Lead the Way Human Factors Are Implemented in Healthcare

Yuval Bitan

CONTENTS

Healthcare working environments and procedures can be improved using human factors methods and tools, but because healthcare is a complex system, it is difficult to implement and evaluate modifications in this environment. If we could test the effects of these modifications in a controlled environment, it would be easier to then implement them on a larger scale in healthcare. The distinct characteristics of paramedic services make them a unique healthcare working environment, which is ideal for implementing and testing such modifications. Paramedic services can, thus, become a good test bed for modifications in a healthcare environment. This chapter will discuss the unique characteristics of paramedic services, describe some of the quality improvement work that was done in paramedic services, and examine what it takes for the paramedic services to lead human factors research and implementation.

INTRODUCTION

This was just another routine call. The pager reported that the patient slipped and fell in his driveway while taking out the garbage. He suffered pain and could not move his leg. It was a dark night and the two paramedics who arrived at the scene crouched around the patient and evaluated his condition. He had severe pain so they started the pain protocol—one dosage of Morphine. One paramedic took a Morphine vial from the medication case and verified that it was the correct vial. Although he was the one who restocked the medication case at the beginning of this shift, and he remembered inserting the Morphine vial into the empty "pocket" in the medication case, he automatically took a look at the label to verify that it was in fact Morphine before opening the vial. His medic partner prepared the IV line while he drew the medication from the vial and got ready to give it to the patient. Soon they will take the patient into the ambulance, he will jump into the driver seat, and they will be on their short way to the hospital. Only at the end of the shift when he and his partner were cleaning the back of the ambulance and restocking the medication case, he saw that one Midazolam vial was missing, while all the Morphine vials were still there. "Which patient got the Midazolam?" he thought to himself. He could not recall any patient that required Midazolam during that shift. Only then did he start to think about the "strange" reaction of the patient with the broken leg to the Morphine. "He was not responding as I would expect from a 40-year-old patient. He shouldn't have knocked unconscious from this dosage." Something went wrong, very wrong. Now the questions started bothering both the paramedics and the patient's family—how could this happen? How could an experienced paramedic like him deliver a dosage of the wrong medication to the patient? There was no "special" pressure during this call, just another routine call, and still such an awful error happened.

Figure 12.1 shows a medication case that paramedics use in some services.

FIGURE 12.1   Medication case.

James Reason (2013) explains that there are many ways in which things can go wrong in any routine process. In this chapter, we will focus on how using human factors tools and methods to implement improvements in healthcare can reduce the chances for such events and how the experience gained from these improvements can serve not only paramedic services but also the entire healthcare system.

HEALTHCARE IS A COMPLEX SYSTEM

The healthcare working environment is known to be dynamic and hectic (Cook and Woods, 1994). It has been compared to complex working environments such as aviation, but there are some characteristics that make the healthcare environment unique even among complex working environments. There are a number of contributors to this complex working environment. In addition, scientific development and new technologies change healthcare in a pace that seems to be accelerating over the years. Understanding the unique characteristics of healthcare is the first step toward developing an approach that may actually generate improvement in this environment.

We should start with the obvious—clinicians need to perform demanding and complicated tasks, some of them life threatening to the patient, under constraints of time and physical location. The need to perform some of these tasks out of the regular working hours is another contributor to the challenge that clinicians face during their work.

Another type of complication is related to the equipment that clinicians use. In most cases, this is unique equipment that was designed and built specifically for clinical care. These technical solutions require the clinicians' technical skills and training. But the lack of standards for the design of many features in medical devices makes their usage challenging and a source of use errors. Another challenge associated with the usage of the equipment is the fact that most healthcare workstations are composed of a collection of devices and instruments which were set ad hoc around the patient bed. This makes each patient room a unique working environment, which clinicians have to familiarize themselves with before starting to work.

Communication gaps between team members with varied clinical backgrounds and experiences are also a source of complexity in healthcare. Among the clinicians who work around the patient bed, the only consistent characteristic is that most of these clinicians will be replaced by their colleagues at shift change, two or three times a day, every day.

However, the main factor contributing to the complexity of this working environment is the conundrum of the human body. The subject of the clinical treatment is a unique organism of which the behavior and responses to interventions are not completely predictable neither by technology nor by experienced professionals. This phenomenon requires the system designers to leave plenty of room for variability and flexibility in any aspect of the healthcare system operation.

IMPROVEMENTS NEED TO HAPPEN

The first signals that patient safety in this complex system might be in danger were identified in the early 1980s (Schreiber et al., 2016) but did not immediately generate much discussion nor action. Clinicians and researchers from within healthcare organizations started to realize that something needs to change, but the early patient safety movement had a slow uptake. Early studies imprinted the areas of research that helped highlight the uniqueness of the healthcare working environment as a complex system that needs to use tools and methods that are similar to those of other high-risk industries. Only in 2000, a series of reports and studies (e.g., Kohn et al.'s [2000] report *To Err Is Human*) alerted the public to the notion that the healthcare complex system is prone to adverse events and these are occurring at an alarming rate. However, this exposure mainly pushed toward the developments of law practice and risk management departments. Only in recent years, we see the evolving of a better understanding of the notion that only design changes that will make the devices, the processes, and the working environment better fit to the user will result in system resilience (e.g., Reason and Reason, 1997). In the last decade, there has been an increase in the research and in the number of consultants working in this area, and more quality improvement projects within the hospitals are focusing on improving patient safety.

WHY IS CHANGE SO SLOW?

One of the characteristics of a complex system is that it is difficult to implement changes. Complex systems are composed of many subsystems with complicated connections and relationships, so it takes a long time until a change in one part affects other parts. In addition, when implementing changes, there is a need for a lot of coordination among the subsystems in order to get the desired result.

Many people believe that implementation of technology and automation processes would improve the healthcare system, but case studies (e.g., Cook and O'Connor, 2005) from many such implementations demonstrate the challenge. Any change in one subsystem affects other subsystems, generating a reaction chain of which the effect on all subsystems is not always predictable. These changes might require new communication and information-sharing channels, customized training for all the staff through their professional groups, and plans for careful transition phase.

Another obstacle for introducing changes is the difficulty in collecting data that will clarify the need for change. There are several factors contributing to this challenge. First, it is not always clear what the goal of the change is. For example, are we trying to improve efficiency or patient outcomes? These are not necessarily "in conflict" but demonstrate the need for clear definition, including the parameters that will measure this modification. Another challenge is that collecting data in the field is a very noisy and imprecise process that introduces a lot of "noise" and variability, which makes it more difficult to draw conclusions.

Lastly, as demonstrated in many reporting system projects initiated by healthcare organizations, healthcare employees have very low incentive to report incidents. The reasons for this are varied, and we will refer here to only two of them. The first reason is the policy to blame and punish (e.g., Leape, 2009) the clinicians at the "sharp end" (nurses and physicians). The second reason is that in healthcare, it is sometimes hard to notice when an adverse event happens, and these incidents are discovered only in retrospect, as what happened in the sample case at the beginning of this chapter. This results in reporting only incidents that are visible (i.e., must be reported) but neglecting all the small incidents that could teach the system a lot (Hollnagel, 2014) but never get to the sunlight.

PARAMEDIC SERVICES: MISSION AND CHALLENGES

Paramedic services (also known as EMS, paramedicine, or prehospital care) are a critical component of public safety and central to the functioning of the healthcare system. Paramedicine usually involves emergencies with very sick patients who need immediate medical attention. Paramedics are often the first point of clinical contact for patients in these critical situations. The challenge faced by paramedics while diagnosing the patients' condition with minimal tools and technology is complicated by the fact that their work is done in places which are not designed for healthcare procedures such as the side of the road. No lab tests or advanced imaging can aid them in diagnosing the patients that, in many cases, are in extreme need for help.

This complex working environment is often fast paced, physically dangerous for staff, and of high stress and requires quick decision-making and action. Interventions performed by paramedics often involve procedures that, if performed incorrectly or at the wrong time, can cause serious harm to patients. In this high-risk environment system, failure can happen, and adverse events occur. Safety incidents such as failure to follow treatment protocol, medication errors, and failed communication have been documented in the literature (Vilke et al., 2007).

Although paramedic services are a critical part in health treatment, it is often a neglected component of the healthcare systems. Historically, paramedic services were affiliated with public safety rather than healthcare. Many services still follow the strict rules of uniforms and ranks that highlight hierarchy in other public safety organizations such as police and firefighters. On the other hand, the field of paramedicine, which is relatively new as a separate discipline, still bases many of its operations on practices inherited from hospital settings, which include the use of devices, procedures, and settings that are similar to the ones used in EDs and intensive care units. Unfortunately, those practices are often inadequate for the unique environment in which paramedics work, with consequences that can ultimately affect the health and safety of both paramedics and their patients.

One problem with paramedic services is the small number of studies that were done on paramedics' work. While other healthcare environments follow the established practice of evidence-based medicine, not many studies have targeted this environment, and the ones that did focused mainly on occupational injuries (e.g., Letendre and Robinson, 2000) and operational aspects (e.g., Giang et al., 2014) rather than clinical and safety issues (e.g., Bigham et al., 2011). One reason for the scarcity of studies might be the fact that the paramedics' working environment is very limited and packed, and it is difficult to conduct research in such an environment (you can find more on the uniqueness of the paramedics' working environment later in this chapter).

UNIQUE CHARACTERISTICS OF PARAMEDIC SERVICES FOR IMPLEMENTING CHANGE

Paramedic services have several unique characteristics that make these organizations better fit for implementing human factors improvements.

SMALL ORGANIZATIONS

Paramedic services tend to be small compared to other healthcare organizations. For example, a city such as Toronto (Canada), with a service area of 650 km² and a daytime population of 3.5 million people, is served by one EMS organization, Toronto Paramedic Services, that employs around 1000 employees. For comparison, the number of hospitals operating in the same city is more than 40, with just one of the hospital networks (University Health Network) that includes 6 hospitals (total of 1200 beds) and has 8000 full-time employees.

FEW PROFESSIONAL GROUPS WITHIN THE ORGANIZATION

As the work in healthcare became more complicated and advanced over the years, healthcare professions became more specialized. These days, healthcare clinicians around the patient bed have many specific proficiencies and specialties. In contrast, the roles of paramedic services have not evolved into many specialized professions and paramedicine was developed into two main levels of expertise—Primary Care Paramedics (also known as EMTs) and Advanced Care Paramedics. In addition, in some jurisdictions, there are also first responders who are part of the paramedic services but have less training and professional experience and intensive care paramedics who handle transport of complicated patients. Still, the majority of the service is provided by just the first two professional groups. This rather simple structure is a major advantage for ease of communication between team members, because they all speak the same "language" and use the same terminology. It is also an advantage for training and developing teamwork.

WORK IN SMALL TEAMS

Paramedics usually work in small teams of two paramedics. In most of these teams, one of the paramedics is at the level of Advanced Care Paramedic, while the other is less senior. This "grouping" dictates how the team splits the work between the two paramedics—based on their accreditations.

Paramedics have limited channels to consult with other clinicians about the clinical treatment that they should provide during the call, and they can rely mainly on their own and their partner's experience. Taking into consideration the importance of collaboration within the paramedic teams while providing care to patients, it is not surprising that although they are not bound to work with the same partner, many paramedics prefer it (e.g., Patterson, 2011) because they find it easier and more efficient.

This type of team is unique to paramedic services since it is very small and the training of both team members is very similar (both Primary Care Paramedics and Advanced Care Paramedics share the same basic training). It guarantees that the communication challenges will be minimal.

LIMITED EQUIPMENT INVENTORY AND WORKING ENVIRONMENTS

The number of equipment items that paramedics use is relatively small. For example, the Ontario Paramedic Services' (Canada) list of equipment for ambulances contains only 115 items, including nonclinical items such as light sticks and towels. An Advanced Life Support ambulance will add 55 items (Provincial Equipment Standards for Ontario Ambulance Services, November 2013). This makes the process of updating and changing the equipment that paramedics use a doable task, if necessary.

Since scenes where paramedics find their patients might be uncomfortable and inappropriate to work, paramedics prefer to take care of their patients in the ambulance. This compact working environment has many limitations, and the access to the patient may be restricted, but it provides a clean and steady work surfaces and good lighting. In contrast to the ambulance, working in the field is known to be challenging not only because of the physical constraints but also due to interruptions from family members and bystanders. This leaves paramedics with just two limited working environments that need to be designed—the paramedics' response bags (also known as jump bags) and the ambulance interior design. Changes in these environments can be done at the level of local services and do not require a big budget for planning and implementing the change.

CLEAR HIERARCHY

As an organization that originated from the public safety service, most paramedic services have a very clear hierarchy and one chain of command. All organizational and clinical decisions are taken under this chain of command, and all members in the organization know their position on this chain. This makes the process of implementing organizational changes much easier than in other healthcare organizations.

SHORT CLINICAL PROTOCOLS

Paramedics' clinical treatment protocols are very clear and relatively short. While the paramedics' main challenge is to identify the most critical clinical conditions that require immediate intervention, once they decide on the clinical protocol, they have to follow that there are not many additional clinical decisions that they should take. Most of the clinical work is technical and based on the detailed protocols that they are well trained to perform. These short protocols allow simple implementation of modifications and improvements within the clinical protocols.

FREQUENT TRAINING SESSIONS

Paramedic services maintain the clinical education of their paramedics through annual training and certification process. Since the equipment that paramedics use is relatively simple, training facilities are within reach for all services and this allows frequent training to all paramedics in the service. While not all countries require paramedics to maintain their certification through annual exams, it is common that all paramedics in the service will go through some kind of training at least once a year.

The mentioned characteristics make paramedic services a domain that can implement change and experiment with new methods that would be more difficult to try in other healthcare domains. This is also a domain that enables easy access to data and simple data collection processes compared to other parts of the healthcare system. We may therefore use paramedic services as a test bed for implementing changes which may inform their application to the rest of the healthcare community.

HUMAN FACTORS PROJECTS IN PARAMEDIC SERVICES

Human factors, which is the study and design of systems, devices, tools, procedures, and processes to better fit users' physical and cognitive abilities, is taking a major part in system improvement. Human factors data collection methods such as observations, interviews, focus groups, and surveys are important for collecting data about the users and their working environment. Tools such as task analysis, heuristics, usability testing, and simulations are used for formative and summative evaluation through the development phase and before and after a change in the current system is designed and implemented. Organizations also use human factors experts to lead processes such as failure mode and effect analysis, root cause analysis, user-centered design, human factors-informed procurement and implementation, and improving teamwork and safety culture evaluation. Following several decades in which human factors took major part in improving safety in other domains, such as aviation, the methods and tools of the human factors have recently begun to change processes, procedures, and working environments in healthcare as well.

The origin of the patient safety movement described earlier in this chapter is based on the understanding that the gap between the working environment and human capabilities is the source of many of the safety incidents in healthcare. This understanding led the way to many projects that involve human factors not only in incident investigation, but also in designing new systems.

As with other elements of change in healthcare, these projects started primarily in hospitals, and although paramedic work is a critical component of the healthcare system, it has not been the recipient of adequate research investments.

Human factors-related projects that were done in paramedic services have mainly targeted the safety of paramedics and patient care. These include, for example, safety culture surveys, human factors-informed procurement, improving teamwork and working procedures, working environment design, medication safety, and more. Some of these projects have been implemented in local services and others in regional services that cover a number of local services. Other projects incorporated larger research and were aimed at implementing global improvements at the level of the entire industry. Local projects are usually smaller in size and scope and are done with local resources and budget, while bigger projects are done by research centers and manufacturers. Figure 12.2 shows a schematic view of how human factors-related projects in paramedic services are covering both the technical and conceptual aspects of paramedics' safety and patient care, while the solutions that they provide can be applied to specific local services or to the global paramedics' community.

Examples of some of the current projects include studies on ways to improve teamwork (Lazzara et al., 2015) and redesign of the paramedic response bags (Bitan et al., 2015) and are changes that could reduce the chances for mistakes in medication, as described in the "Introduction." Figure 12.3 demonstrates new paramedic response bags that were developed in Northumberland (Ontario, Canada). Other projects target the ambulance interior design, which has not changed much since World War I. Current projects to design the future ambulance apply user-centered design tools to redesign the back of the ambulance in a way that will better fit the paramedics' needs. These projects are done both as research projects to test new concepts (such as Redesigning the Emergency Ambulance from the Helen Hamlyn Centre for Design, Royal College of Art) and as a commercial solution such as Ferno 2020. Projects that aim to improve the technical part of the work paramedics do bring new equipment that carries patients (such as chairs and stretchers). These projects focus the design on technical solutions that will reduce the paramedic physical burden and add features that are more ergonomically and safety designed. Many of these projects are still at a stage of research and have not yet been implemented, but when they are implemented in paramedic services, this may also facilitate their implementation into other areas in healthcare, which may benefit from the lessons learned in this smaller-scale modification.

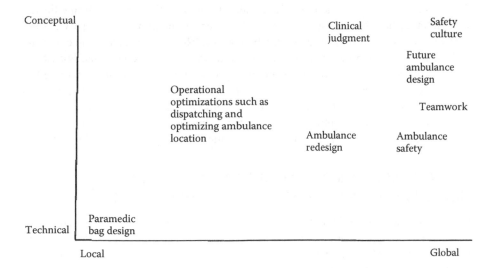

FIGURE 12.2   Level of human factors projects in paramedic services.

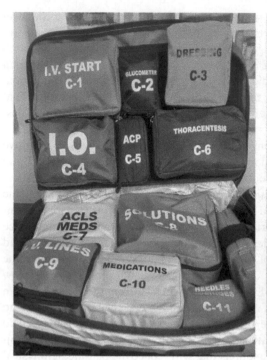

FIGURE 12.3   New paramedics' response bags, a project that was done in Northumberland (Ontario, Canada).

Despite the fact that paramedic services and other healthcare systems are different in many ways, there are important similarities across both systems, which enable the transfer of knowledge acquired in implementing modifications in paramedic services to other healthcare systems. Although projects done in paramedic services are typically of a smaller scale, lessons can be learned from these projects and can teach us how to use human factors tools and methods in larger-scale systems. For example, starting a quality improvement project by working with a group of paramedics to study the way that things currently work in the field, design the recommended improvement, and carefully plan the implementation stage can teach us a lot about the challenges of implementing such projects in a hospital. Learning from the experience of the paramedic services will serve as a pilot for a larger-scale implementation.

WHAT IT TAKES FOR PARAMEDIC SERVICES TO LEAD HUMAN FACTORS IMPLEMENTATION IN HEALTHCARE

In order to become leaders in implementing human factors tools and methods to improve patient safety, paramedic services should embrace discovery and innovation. Programs that encourage paramedics to initiate and participate in safety improvement and research projects should be promoted. Such changes should begin from the level of the management that has to show leadership by facilitating and encouraging these projects. This management focus will lead to a culture shift toward embracing change and innovation even at the cost of difficulties along the path.

Improvements and modifications of equipment and processes cannot be designed by outsiders. One of the key features for success in such projects is getting the paramedics involved in the process. The paramedics will inform the process by providing valuable information from their own experience in the field and participating in interviews and observations. Importantly, their involvement in the design phase will increase the compliance during the implementation phase (Bitan et al., 2015). One element which can promote this initiative is including a research component as part of the paramedic training requirements. This goes in line with the fact that many paramedics receive their training through college degrees. Including research as part of the paramedic college studies would introduce them to concepts of evidence-based improvements and will develop a working environment of professionals who plan change and act upon it based on collected data. One group that had a significant impact on the implementation of human factors in healthcare more generally was that of anesthesiologists, who were seeking solutions for the technical challenges that they faced in their work. Paramedics can take a similar role in leading human factors change in healthcare. This will include not only being actively involved in projects but also participating in discussions, being present in conferences, and publishing in professional journals. By becoming a leading force in human factors projects in healthcare, paramedic services not only will have the benefit of developing a safer working environment but will also change the safety culture in this community to become a community where strive for change and improvement is part of its DNA.

SUMMARY

Paramedic service is a domain that has unique characteristics among healthcare organizations. These characteristics allow paramedic services to implement improvements faster and more easily. These improvements can reduce the chances for safety incidents such as the one described in the "Introduction." In addition, sharing the experience of such projects with other healthcare organizations will make the paramedic services a leading force in this complex domain.

Paul Raftis, the Toronto Paramedic Services chief, highlighted the change in the paramedic services world: "Today's paramedics are college graduates, highly educated and highly trained in pre-hospital paramedicine that is life-changing and life-saving. These paramedics use modern, state-of-the-art technology and clinical best practices, to help our patients when they need it most. These changes have evolved the Paramedic Service so significantly since the 1970s that the two are barely recognizable when compared to each other" (Toronto Paramedic Services, June 2015). Adding research and quality improvement projects as one of the areas where paramedic services can develop will benefit the paramedic services and will allow them to become a test bed for implementing human factors in the healthcare system.

REFERENCES

Bigham, B.L., Buick, J.E., Brooks, S.C., Morrison, M., Shojania, K.G. & Morrison, L.J. (2011). Patient safety in emergency medical services: A systematic review of the literature. *Prehospital Emergency Care, 16*(1), 20–35.

Bitan, Y., Ramey, S., Philp, G. & Uukkivi, T. (2015, June). Working with paramedics on implementing human factors improvements to their response bags. In *Proceedings of the International Symposium on Human Factors and Ergonomics in Health Care, 4*(1), 179–181. Thousand Oaks, CA: Sage Publications.

Cook, R.I. & O'Connor, M.F. (2005). Thinking about accidents and systems. In H.R. Manasse Jr. & K.K. Thompson (Eds.), *Medication safety: A guide for healthcare facilities* (pp. 73–88). Bethesda, MD: American Society of Health Systems Pharmacists.

Cook, R.I. & Woods, D.D. (1994). Operating at the sharp end: The complexity of human error. In M.S. Bogner (Ed.), *Human error in medicine* (Vol. 13, pp. 225–310). Hillsdale, NJ: Erlbaum Associates.

Giang, W.C., Donmez, B., Fatahi, A., Ahghari, M. & MacDonald, R.D. (2014). Supporting air versus ground vehicle decisions for interfacility medical transport using historical data. *IEEE Transactions on Human–Machine Systems, 44*(1), 55–65.

Hollnagel, E. (2014). *Safety-I and safety-II: The past and future of safety management.* Burlington, VA: Ashgate.

Kohn, L.T., Corrigan, J.M. & Donaldson, M.S. (Eds.). (2000). *To err is human: Building a safer health system* (Vol. 6). Washington, DC: National Academies Press.

Lazzara, E.H., Keebler, J.R., Shuffler, M.L., Patzer, B., Smith, D.C. & Misasi, P. (2015). Considerations for multiteam systems in emergency medical services. *Journal of Patient Safety.* DOI: 10.1097/PTS.0000000000000213.

Leape, L.L. (2009). Errors in medicine. *Clinica Chimica Acta, 404*(1), 2–5.

Letendre, J. & Robinson, D. (2000). *Evaluation of paramedic's tasks and equipment to control the risk of musculoskeletal injury.* Ambulance Paramedics of British Columbia. CUPE Local 873 Internal Report 6-08-0793.

Patterson, P.D., Weaver, M.D., Weaver, S.J., Rosen, M.A., Todorova, G., Weingart, L.R. et al. (2011). Measuring teamwork and conflict among emergency medical technician personnel. *Prehospital Emergency Care, 16*(1), 98–108.

Reason, J. (2013). *A life in error: From little slips to big disasters.* Farnham, UK: Ashgate.

Reason, J.T. & Reason, J.T. (1997). *Managing the risks of organizational accidents* (Vol. 6). Aldershot, UK: Ashgate.

Schreiber, M., Klingelhöfer, D., Groneberg, D.A. & Brüggmann, D. (2016). Patient safety: The landscape of the global research output and gender distribution. *British Medical Journal Open, 6*(2), e008322.

Toronto Paramedic Services. (2015). Paul Raftis, Chief, Toronto Paramedic Services (http://torontoparamedicservices.ca/). Retrieved from https://youtu.be/WP28A9vh_ZM.

Vilke, G.M., Tornabene, S.V., Stepanski, B., Shipp, H.E., Ray, L.U., Metz, M.A. et al. (2007). Paramedic self-reported medication errors. *Prehospital Emergency Care, 11*(1), 80–84.

13 Disaster and Response in Emergency Medical Services
Team Training to the Rescue*

*Shirley C. Sonesh, Megan E. Gregory,
Ashley M. Hughes, and Eduardo Salas*

CONTENTS

INTRODUCTION

On April 20, 1999, Emergency Medical Service (EMS) providers in Littleton, Colorado responded to Columbine High School after reports of a mass casualty situation. Upon arrival, responders noted an active, violent, ongoing incident, with numerous casualties, an unknown number of heavily-armed assailants whose whereabouts were unknown, fleeing students and staff, and reports of explosions, odors of natural gas, and an activated fire alarm (Mell & Sztajnkrycer, 2004).

---

* The views expressed in this work are solely those of the authors and not those of the Florida Department of Health. The views expressed in this chapter are those of the authors and do not necessarily reflect the position or policy of the Department of Veterans Affairs or the US government.

In this situation, first responders from various agencies and divisions had to work together to triage and treat patients both inside and outside of the school within this active and dangerous scene. Flawless execution of teamwork, defined as the necessary attitudinal, behavioral, and cognitive mechanisms through which team members accomplish shared goals (Salas, Burke, & Cannon-Bowers, 2000), is essential to execute tasks in high-stakes situations such as disaster management. Due to their training and experience, EMS providers are skilled in providing high-quality care to stabilize patients during disaster events. However, although teams of EMS providers may be composed of experts, this does not mean that they are necessarily an "expert team" (Salas et al., 1997). That is, in order to function effectively as a team, EMS providers must not only possess the task-relevant skills, but also be trained to develop and use team skills.

The purpose of this chapter is to define and contextualize what teamwork looks like in the EMS setting in order to describe the anticipated benefits of team training in EMS and particularly in disaster management. Furthermore, this chapter will leverage and discuss the science of team training to pave a way forward for EMS to adopt, use, and sustain teamwork skills that serve to facilitate disaster response. Specifically, our aims are to (1) identify and explicate the team competencies necessary for superior EMS patient care, management, and regulation during disaster situations and (2) leverage the science of team training to propose a way forward for the promotion, implementation, and sustainment of teamwork in EMS disaster management.

WHAT IS DISASTER MANAGEMENT?

Disaster management situations are defined as events that overwhelm the capabilities and resources of the local emergency response system (Quarentelli, 1985), including events such as devastating tornadoes, mass shootings, and hurricanes, among others. EMS disaster response, in particular, is worthy of focused examination as it is characterized by high stakes, many stressors, and the heightened need for coordination. The conditions surrounding the event and response efforts can vary greatly in terms of ongoing danger, availability of resources, and number and type of responders. Thus, there is no one pattern or formula that a disaster management situation falls under. Oftentimes, the EMS system must rely on the coordination of EMS providers, as well as other EMS teams, to effectively manage emergency response and disaster management situations.

Moreover, disasters often unfold rapidly and unexpectedly and result in a wide variety of patient injuries, requiring immediate medical care. This care is often difficult to provide due to failing infrastructure and destroyed and/or exhausted resources that impede the delivery of EMS services to the impacted individuals (Cardenas & Daccord, 2013). Additionally, disaster and mass casualty events require vigilance and recognition of subtle trends in patient symptoms. If there is improper triaging or missed injuries, the need for repeated operations can occur. Efficient teamwork can potentially overcome such obstacles and result in more accurate diagnoses and treatment plan development. Throughout this chapter, we will contextualize our discussion toward the disaster response setting, highlighting specific needs and areas of concern in these situations.

TEAMWORK IN EMS DISASTER MANAGEMENT

Teams consist of two or more individuals who share a commitment to common goals and are part of a larger organizational system in which each team member has a different skill set, role, and responsibility (Salas et al., 1992). Teams often must interact among themselves and with external members in order to be effective (McGrath, 1984). Such interaction is necessary in order to appropriately make decisions (Hollenbeck et al., 1995), coordinate to accomplish interdependent tasks (Saavedra et al., 1993), and achieve shared goals (Aubé & Rousseau, 2005). Successful team interaction requires effective teamwork. As previously stated, *teamwork* is defined as the attitudinal, behavioral, and cognitive mechanisms through which team members accomplish shared goals (Salas et al., 2000). Teamwork impacts performance and goal achievement and, in clinical contexts, is associated with patient outcome improvement (McCulloch et al., 2009; Capella et al., 2010; Gregory et al., 2014), quality of patient care (Leonard et al., 2004), and patient safety (Patterson et al., 2012).

While critically important, there has been little focus on teamwork breakdowns in the EMS literature (Williams et al., 1999; Hughes et al., 2013). Adverse outcomes that occur in EMS often place too much blame on individual- or system-level problems and typically fail to consider the critical influence of teamwork. Despite teamwork being a major part of the medical delivery system and critical for ensuring patient safety (Joint Commission, 2014), teamwork competencies are not commonly incorporated into EMS training or quality improvement efforts. This is problematic, since preventable patient harm is somewhat common in prehospital patient care. For example, medication dosing errors alone have been found to account for 15% of errors resulting in adverse events in prehospital care (Fairbanks et al., 2008). Furthermore, medication and other medical error events are estimated to be grossly underreported (Nguyen & Nguyen, 2008). As previously stated, teamwork is vital for EMS providers to work together safely to provide patient care (Vilensky, 2011). Particularly, in instances of disaster management, EMS providers must rely on each other, communicate, and provide backup behavior to mitigate the potential effects of fatigue and burnout inherent to overwhelming conditions (Weaver et al., 2012) in order to mitigate error.

However, EMS as an industry often fails to perceive their work as being team oriented (Patterson et al., 2011a), and as such, EMS providers do not typically consider themselves to be a team, thus impacting performance and patient care. Influencing factors noted in the literature indicate that this may be because they often work in dyads, with one provider driving the vehicle and the other caring for the patient (Patterson et al., 2011a). Furthermore, evidence suggests that EMS team members question each other's skills, integrity, and patient empathy, which stunts the development of positive teamwork behaviors such as trust, backup behaviors, and closed-loop communication (Brown et al., 2005; Patterson et al., 2013). Additionally, providers are often somewhat unfamiliar with each other; data suggest that providers work with an average of 19.3 partners annually (Patterson et al., 2011a), making it difficult to develop a perception of intact team membership. One of the consequences of this is that partners tend to provide fewer backup behaviors when they are unfamiliar with each other (Smith-Jentsch et al., 2009), which, in a fast-paced, error-prone, stress-inducing environment such as disaster management, compromises the adequacy of patient care, patient safety, and provider safety from injury (Hughes et al., in press). In fact, lack of team familiarity has been shown to hurt performance as it accounts for adverse outcomes in EMS (Bayley et al., 2007; Eschmann et al., 2010). While there are many reasons that teamwork is not perceived to be a necessary component for effective care in EMS, EMS providers working together *are* a team, as they are working interdependently toward a shared goal-effective patient care. In this section, we will describe the critical teamwork factors that are necessary for any high-stakes team performance and contextualize them to the EMS and disaster management context.

THE SEVEN CS OF TEAMWORK

Salas et al. (2014) stipulate that the critical components of effective teamwork for any type of team can be summarized by a heuristic that they call *the seven Cs*. These Cs include communication, cooperation, coaching (i.e., leadership and mentoring), conflict, coordination, cognition, and conditions (Table 13.1). In this section, we will define and discuss each of these Cs in the context of EMS disaster management. In doing so, we will also leverage nine components identified by Patterson et al. (2011b) as specific to teamwork in the EMS setting. The nine components are (1) team orientation; (2) team structure and leadership; (3) partner communication, team support, and monitoring; (4) partner trust and shared mental models; (5) partner adaptability and backup behavior; (6) process conflict; (7) strong task conflict; (8) mild task conflict; and (9) interpersonal conflict. Each of these can be mapped onto one of the seven Cs. Throughout this section, we will make note of instances where this is the case. This will allow for a comparison of team competencies described by the broader-context general team literature to those that are specific to the EMS domain.

Communication is defined as a reciprocal process whereby two or more team members send and receive information (Salas et al., 2014). Patterson et al.'s (2011b) EMS teamwork scale emphasizes the role of clear communication and its effect on team support and monitoring, as communication plays an integral role in many key team processes (Wahr et al., 2013). High-quality communication is essential in EMS teams, as poor communication is frequently cited as a root cause of medical error (Joint Commission, 2014). Furthermore, meta-analytic evidence suggests that information sharing within teams is related to team performance (Mesmer-Magnus & DeChurch, 2009). The research is clear: high-quality communication is characterized by communication that is closed loop (i.e., team members' assurance that communication was received and understood correctly), clear, and open (Ellingson, 2002; Salas et al., 2008c; Mesmer-Magnus & DeChurch, 2009). In disaster management, open communication is particularly important, as the rapidly changing environment may enhance conditions that facilitate different team members having unique pieces of information that are essential to full and complete understanding

TABLE 13.1

The Seven Cs for EMS Disaster Response

The Seven Cs of Teamwork (Salas et al., 2014)	Definition	Applicability to EMS Disaster Response
Communication	Sending and receiving information between team members	42% of air medical transfer calls contained a communication error (Vilensky & MacDonald, 2011).
Cooperation	Motivational drivers; the attitudes, beliefs, and emotions of a team	"It's a sign of weakness if you ask another paramedic on the scene 'what do you think?' "—interviewee from Fairbanks et al. (2008, p. 637), demonstrating a climate of low psychological safety.
Coaching	Engaging in leadership behaviors in order to accomplish team goals	68% of sentinel events in medicine have poor leadership as a root cause (Joint Commission, 2014).
Conflict	Perceived incompatibilities between two or more team members	Work conflict is negatively related to morale and job satisfaction in firefighters and paramedics (Beaton et al., 1997).
Coordination	Engaging in behaviors and cognitions necessary to perform team tasks	Task complexity (multiple, concurrent tasks; uncertainty; changing plans; high workload) in EMS makes team coordination especially difficult (Xiao et al., 1996).
Cognition	Shared understanding among team members about roles, goals, norms, and team member's KSAs	Flin and Maran (2004) observed acute medicine teams and found that when they did not work well together, there had always been a breakdown in team cognition at some point.
Conditions	Includes the culture (i.e., values, beliefs, and norms within an organization or environment), composition (i.e., the configuration of team member's KSAs and diversity within a team), and context (i.e., situational factors that impact behavior and outcomes)	EMS personnel who experience fatigue are less likely to rely on their partners for support (Patterson et al., 2013).

of a patient's condition. Therefore, it is vital that team members communicate any unique information to the entire team. Research has found that groups tend to do a poor job at communicating unique information (Stasser & Titus, 1985; Wittenbaum & Stasser, 1996); however, doing so improves decision quality (Winquist & Larson, 1998). Team training for EMS in disaster management should aim to train team members to probe for unique information, an endeavor which has been successful in prior research (Larson et al., 1994; Wittenbaum, 1998, 2000). Doing so will likely yield more open communication, improved patient care, fewer errors, and more accurate triage of patients in disaster situations.

Cooperation is viewed by Salas et al. (2014) as a motivational driver of teams. Specifically, cooperation involves team member attitudes, beliefs, and emotions that drive behavior. Trust, defined as "the intention to accept vulnerability to a trustee based on positive expectations of his or her actions" (Colquitt et al., 2007, p. 909), is particularly relevant for EMS teams in disaster management situations. In particular, trust leads to higher task performance and more risk-taking behaviors (Colquitt et al., 2007), both are necessary in disaster management situations. Psychological safety, another aspect of cooperation, is described as "a shared belief that the team is safe for interpersonal risk taking" (Edmondson, 1999, p. 354). Psychological safety is essential in facilitating open communication, as a team characterized by high psychological safety is more likely to have members who speak up (Edmondson, 1999). Furthermore, the team orientation construct as described by Patterson et al. (2011b) falls under the umbrella of cooperation. This construct involves one's orientation or openness to being a member of a team, in comparison to working alone. Similarly, Patterson et al. (2011b) would likely categorize part of their construct of partner trust and shared mental models, as cooperation, as it involves trust of other team members' knowledge, skills, and abilities (KSAs).

Coaching involves "the enactment of leadership behaviors to establish goals and set direction that leads to the successful accomplishment of these goals" (Salas et al., 2014, p. 5). These leadership behaviors can involve motivating the team, establishing expectations, and monitoring team progress and the environment. Coaching can be enacted by any member of the team, regardless of whether or not they are recognized as a formal leader (Pearce et al., 2009; Morgeson et al., 2010). In fact, distributing leadership among members in this way can prevent one team member from becoming overloaded and can improve team performance (Carson et al., 2007). This bears similarity to the team structure and leadership teamwork competency as described by Patterson et al. (2011b), which is defined as understanding who is undertaking which leadership roles. Coaching leads to positive outcomes such as higher team performance, team effectiveness, team learning, and job satisfaction (Burke et al., 2006; Barling et al., 2011; DeRue et al., 2011). Coaching should be performed throughout all phases of a team's life cycle (Kozlowski et al., 2009), including the transition (i.e., planning), action (i.e., task completion), and interpersonal (i.e., conflict management) phases as delineated by Marks et al. (2001). In disaster management situations, coaching behaviors such as monitoring progress and the environment become especially important. Because these situations are often characterized by multiple patients and quickly changing and uncertain circumstances, EMS team members must be certain to continuously assess what is going on with each patient and the environment around them.

Conflict is defined as the "perceptions by the parties involved that they hold discrepant views or have interpersonal incompatibilities" (Jehn, 1995, p. 257). Conflict is considered to have different subtypes (see Jehn, 1994, 1995, 1997), including relationship conflict (i.e., interpersonal issues among team members) and task conflict (i.e., disagreements about the task). Similar to Jehn's (1994, 1995, 1997) conceptualization, Patterson et al.'s (2011b) measure of EMS teamwork describes different types of conflict, including mild and strong task conflict, process conflict (conflict about the way in which tasks are completed), and interpersonal (relationship) conflict. Meta-analytic evidence suggests that both task and relationship conflicts are strongly and negatively related to team performance and team member satisfaction (De Dreu & Weingart, 2003). However, other evidence suggests that a moderate amount of task conflict may be beneficial for team performance in the absence of relationship conflict (Shaw et al., 2011). In either case, the appropriate management of conflict is vital to successful performance in disaster management situations. Conflict among providers has been cited as a problem inherent within the EMS settings (Patterson et al., 2012) and has been cited as being a contributor to medication errors (Sonesh et al., 2013). In disaster management response, it is likely that teams will encounter unclear and novel situations and conditions that have no clear-cut, correct approach. EMS team members must be willing to see all viewpoints when task conflict occurs during these tense situations. Furthermore, team members must not let preexisting or current relationship conflict cause harm to patients.

Coordination is defined as the enactment of behaviors and cognitions in order to perform team tasks and transform team resources into outcomes (Sims & Salas, 2007; Salas et al., 2014). Many behaviors can be characterized as coordination, so long as their intent is to orchestrate "the sequence and timing of interdependent actions" (Marks et al., 2001, p. 363). Patterson et al.'s (2011b) partner adaptability and backup behavior construct falls into the category of coordination, as it entails adapting to changing situations and anticipating other team members' needs for help. Coordination can be explicit, wherein team members purposefully work together to act interdependently. Alternatively, coordination can be implicit, such that team members can act interdependently without communication or instruction due to a shared understanding that allows team members to anticipate others' needs and behaviors (Rico et al., 2008).

When considering coordination efforts, it is essential to first define the level of interdependence that a task requires. Saavedra et al. (1993) define various levels of interdependence, ranging from pooled tasks that do not require direction interaction among team members, to sequential tasks requiring one team member to act before another one can act (e.g., an assembly line), to more complex types of interdependence such as reciprocal and team interdependence. Tasks requiring reciprocal interdependence (e.g., stabilizing critical patients) involve some structure in regard to team roles; however, these types of tasks allow for two-way interactions among team members and flexibility in regard to the order in which parts of the task occur. Team interdependence is even more unstructured and involves team members engaging in a complex pattern of mutual interactions. The level of interdependence has implications for coordination. In a task requiring sequential interdependence, for example, coordination will require an awareness of the progress of each team member's task. In tasks requiring more complex levels of interdependence, coordination can be more of an emergent process that evolves over the course of task completion.

Coordination within teams has been linked to higher team performance (Stewart, 2006). In disaster management situations, coordination among the numerous members of the team is essential in order to prevent unnecessary duplication of efforts, which wastes valuable resources. Furthermore, coordination in these settings ensures that all required plans and actions have been carried out, an important feature in hectic, fast-paced disaster management situations.

Cognition refers to a shared understanding among team members in regard to roles, responsibilities, goals, norms, the situation, and the KSAs of each team member (Wildman et al., 2012). Meta-analytic evidence suggests that team cognition is related to team performance (DeChurch & Mesmer-Magnus, 2010). Patterson et al. (2011b) would likely consider the construct of partner trust and shared mental models to fall under the cognition umbrella, as this construct considers the extent to which team members are aware of other team members' KSAs, strengths, and weaknesses. In order to obtain shared cognition in a disaster management situation, EMS teams should take a moment at the onset of the response effort to brief. In other words, the team should get on the same page in regard to what roles and responsibilities that each team member will undertake, what the specifics of the situation are, and what the goals and objectives for the response are. The momentary time taken out to engage in this process will likely save time by mitigating potential confusion and duplication of efforts throughout the disaster management response period.

Lastly, *conditions*, according to Salas et al. (2014), include culture (i.e., values, beliefs, and norms within an organization or environment), team composition (i.e., the configuration of team members' KSAs and diversity within a team), and context (i.e., situational factors that impact behavior and outcomes). Salas et al. (2014) stipulate that conditions facilitate or hinder effective teamwork. In EMS settings, organizational culture is a condition particularly deserving of attention. Organizational culture has been defined as a "pattern of basic assumptions" which influences the norms for behavior, attitudes, and cognitions within organizational settings (Schein, 1985, p. 3). The culture of medicine is one that values flawless technical skill (Oriol, 2006), such that education for clinicians is typically done in silo by discipline or profession (Harden, 2006). Thereby, with focus on individual skills, enactment of teamwork fails to receive recognition. While the hierarchy permeating medical culture has been examined in detail within traditional medical settings such as hospitals, the hierarchy within EMS settings has been noted as a barrier to effective disaster management response (Auf der Heide, 2006). A prominent example of how norms influence teamwork is a situation in which a junior EMS practitioner does not speak up during near-miss or error events committed by a more senior EMS practitioner (Edmondson, 2003). In a disaster situation, such an organizational culture may prevent personnel from requesting help from team members, which may increase the likelihood of a medical error in an already chaotic disaster situation (Fairbanks et al., 2008).

While culture permeates the influencing factors in EMS behaviors, there are several additional influencing conditions that are inherent to the EMS system. One of the most salient conditions of the EMS work environment is that it is characterized by multiple stressors (Lammers et al., 2012). Specifically, the conditions under which EMS providers experience high stress and anxiety due to high patient load, which presents unique coordination demands and negatively impacts teamwork (Bentley et al., 2013). EMS agencies are reported to respond to as many as 1700 calls per day, and consequently, EMS providers are delayed in responding to new incoming calls (Peleg & Pliskin, 2004), creating work overload. Similarly, the immediacy and number of patient calls in a disaster situation result in high cognitive workload, as well as physical workload (Meshkati & Hancock, 2011). High workload affects one's ability to process and filter critical information (Wickens & Hollands, 2000), which is essential in EMS disaster response. Furthermore, EMS providers work a 48-hour on and 24-hour off schedule, which has been cited as a source of fatigue (Beaton & Murphy, 1993) and linked to human performance issues. Fatigue is defined as a cognitive and physical state in which perceived exertion exceeds individual capacity to complete a task (Noakes et al., 2005) and may pose a challenge in enacting effective teamwork behaviors during disaster response. To elaborate, there is evidence that fatigue reduces motivation (Boksem et al., 2006) and has been linked to severe decreases in patient safety in the EMS context (Ulmer et al., 2009). In fact, EMS personnel who experience fatigue are less likely to rely on their partners for support (Patterson et al., 2013). Shockingly, a study by Patterson et al. (2010) found that 44.5% of EMS personnel in their study experienced severe fatigue. During a disaster management scenario, fatigue is likely to occur quickly due to the length and intensity of these situations, and team performance is likely to suffer as a result. EMS providers are attempting to work on tasks that are largely ill-defined, within the context of multiteam systems (e.g., EMS, law enforcement, special weapons and tactics), and the number of casualties often outnumbers the resources of the EMS organizations. These influencing conditions are projected to exacerbate decrements in team performance in a disaster management situation.

A WAY FORWARD: TEAMWORK TRAINING IN EMS DISASTER MANAGEMENT

In emergency medical care, response time is critical (De Bruycker et al., 1983, 1985). Not only are medical disaster response models and operation plans necessary (Schultz et al., 1996), but so are the effective communication and interteam and intrateam coordination within and between EMS teams. As alluded to previously, the inherent interdependencies between teams during a disaster situation require increased coordination and create a need to exchange information, share knowledge, and resolve emerging conflict (Galbraith, 1973) both within and between teams. To overcome and adequately respond to disaster situations, teamwork within and among EMS teams are critical, as it serves as a form of checks and balances in EMS partnerships (Rosen et al., 2008). These competencies can be trained and applied to the job by adopting a team training curriculum that has shown to be effective in improving individual, organizational, and patient outcomes in other fields of healthcare (Gregory et al., 2014).

As these teamwork behaviors are critically important to the provision of safe and efficient healthcare in the hospital setting (Neily et al., 2010), hospitals have begun to adopt team training initiatives to impart team knowledge, skills, and attitudes onto their healthcare providers (Weaver et al., 2010a). However, while the field of EMS suffers from similar consequences of poor teamwork (Williams et al., 1999), it has typically failed to adopt initiatives to promote, reinforce, and sustain teamwork among EMS partners in the field, despite calls for this field to do so (Williams et al., 1999). EMS providers are prone to the same types of behavioral and cognitive errors as healthcare providers in other settings; unfortunately, there is little research in this domain aimed at resolving teamwork issues in disaster situations. Nonetheless, EMS researchers recommend teamwork and leadership training for all new EMS personnel (Tomek, 2008; Pianezza, 2010; Patterson et al., 2013). As Pianezza (2010) puts it, "In EMS, leadership is traditionally given to those with seniority. While this is not always a bad idea, without correct training and mentoring, that person will fail."

Team training is defined as "a set of instructional strategies and tools aimed at enhancing teamwork knowledge, skills, processes, and performance" (Tannenbaum et al., 1996, p. 516). Team training initiatives have been shown to improve not only team performance (Salas et al., 2008a) but also team processes (e.g., communication, coordination, shared goals), which are crucial for EMS teams providing patient care. In fact, studies examining teamwork interventions found that those who received team training had decreased symptoms of burnout and increased self-efficacy in stressful medical settings (Buljac-Samardzic et al., 2010). Nonetheless, the institutionalization of team training across the EMS stetting must continue to be addressed. While the healthcare community at large has made considerable progress in designing and implementing team training across a number of settings, the EMS setting still trails behind. In fact, in a recent qualitative review examining the effectiveness of teamwork training in healthcare contexts, there were no instances of team training in EMS (Weaver et al., 2010a). This scarcity of evidence for team training as it impacts EMS performance highlights the need for EMS as an industry to adopt teamwork training initiatives.

The EMS context has been called "the least ideal physical and emotional environment, creating a milieu ripe for patient harm" (Bigham et al., 2011, p. 20). While highlighting a need for improved teamwork in this setting, this quote also implies that the EMS setting is a particularly difficult one in which to train teamwork competencies. While many have noted the importance of teamwork in EMS (Williams et al., 1999; Vernon, 2013), the use of team training in this context is not yet widely leveraged. For this reason, we sought to delineate a clear path forward for EMS practitioners to take to promote, implement, and sustain the teamwork competencies necessary to achieve optimal performance in disaster management situations. We organize the "way forward" by leveraging the science of team training (Salas et al., 1997) to discuss what EMS agencies can do *before*, *during*, and *after* team training in order to create an environment that is ready for and supportive of teamwork training initiatives (Gregory et al., 2013) (Table 13.2).

BEFORE TEAM TRAINING

Perhaps, the most important part of a team training program occurs before the training even takes place. It is essential to determine what to train, as well as to obtain buy-in from trainees. In this section, we will briefly discuss how to achieve these goals in the context of team training for EMS providers in disaster management situations.

Need Analysis

First, team training should be systematically designed as effective training should identify specific needs using a training need analysis (Goldstein & Gilliam, 1990). Conducting a team training need analysis can better inform the teamwork

TABLE 13.2

Before, During, and After EMS Team Training

Training Phase	Tools/Mechanisms/Activities	Source
Before training	*Conduct need analysis:* conduct person, organizational, and team task analyses to shape the team training program	Goldstein (1991)
	Obtain buy-in: motivate trainees by explaining how teamwork affects their work	Kotter (1995)
During training	*Choose an instructional strategy and training method:* provide enough opportunities to practice in a disaster management context	Weaver et al. (2010b)
	Leverage SBTT: script scenarios that target a subset of team competencies using triggers that elicit team behavior	Weaver et al. (2010b); Owen et al. (2006)
	Debrief/provide feedback: provide supportive environment, structured, developmental, and specific feedback on both good and bad examples of teamwork	Salas et al. (2008b)
After training	*Evaluate:* evaluate training at multiple levels: reactions, learning, behaviors, and results	Kirkpatrick (1959)
	Reinforce: provide support, incentives, and practical tools	Grossman and Salas (2011)
	Sustain: strengthen culture of teamwork and conduct periodic refresher training	King and Harden (2013)

competencies necessary for EMS teams to be trained on to optimally perform in a disaster management situation (see Bowers et al., 1998). For team training, one should conduct analyses of the organization, the trainees, and the teamwork tasks performed on the job. The *organizational analysis* involves determining the extent to which the organization supports the training, by providing resources and encouraging employees to participate in the training. If the organization or supervisors do not support the training, it will not be successful, as trainees will have little motivation to engage. Furthermore, understanding the resources that the organization has will help determine decisions such as what training strategy (e.g., information, demonstration, practice) is feasible (Salas & Cannon-Bowers, 2001). A *person analysis* also needs to be conducted to determine the characteristics of potential trainees. For teamwork training, constructs such as collective orientation (i.e., the extent to which one prefers to work in a team, as compared to their preference for working alone) and collective efficacy (i.e., the extent to which one believes that their team will be successful) are particularly relevant. Lastly, a *team task* analysis (Bowers et al., 1998) involves identifying the tasks that employees perform that require teamwork. By knowing this, one can dig deeper into the teamwork competencies (e.g., leadership, mutual support.) that may be of the highest importance for the employees. As previously mentioned, Patterson et al. (2011b) conducted a survey regarding this information for EMS providers. According to these data, the teamwork competencies most salient in this industry (and, thus, potentially most essential for training) include interpersonal conflict, partner trust and shared mental models, team structure and leadership, process conflict, and strong task conflict. However, it is likely that the team needs will vary somewhat under the conditions of a disaster situation: multiple patients, uncertain and ongoing danger, and overwhelmed resources. Thus, it is important to consider the unique context of disaster situations and, as such, conduct a separate need analysis for this purpose.

The time pressure, fluidity of EMS teams, and ambiguity present in disaster situations make it necessary to identify and target who will receive team training. It is important to be as inclusive as possible in deciding who receives team training, as even the smallest contributor to the team has a huge impact on patient outcomes and on the successful management of patients. Moreover, the skill level of the trainees must be taken into account. EMS practitioners are inherently multidisciplinary, as they represent different backgrounds, levels of expertise, skill sets, and experiences.

Obtain Buy-In

It is important to obtain buy-in from trainees before beginning the team training program. That is, for EMS to adopt teamwork as a norm, teamwork must become integrated into the fabric of EMS culture (King & Harden, 2013), and this requires the recognition of the urgent need for teamwork (Kotter, 1995). This will increase trainee motivation, which has been associated with higher levels of learning, retention, and transferring learned competencies to the workplace (Salas & Cannon-Bowers, 2001). One way to accomplish stakeholder buy-in in this context is to demonstrate that team training is effective at improving the coordinating mechanisms inherent to disaster management. The empirical evidence of team training's effectiveness is emerging; in fact, a meta-analysis on team training in various high-stakes industries demonstrates that team training significantly improves team attitudes, behaviors, and cognitions (Salas et al., 2008a). Furthermore, recent meta-analytic evidence demonstrates the effectiveness of team training across healthcare domains (Gregory et al., 2014; O'Dea et al., 2014; Hughes et al., 2016). The evidence is emerging that team training works, yet demonstration of this and key principles of team training need to be implemented and evaluated within EMS contexts. By demonstrating the success of team training initiatives within EMS agencies, providers can view the impact of team training first hand, facilitating buy-in through increased perceptions of utility, which are integral to adoption of team training practices (Axtell et al., 1997; Armenakis et al., 2000).

This empirical evidence geared toward improving perceptions of team training's utility should be coupled with anecdotal success stories of training implementation to gain buy-in at all levels of the agency and organization (e.g., supervisors, top management, frontline providers, support staff). One way to motivate trainees is to begin the training with a definition of what a team is and clarify how EMS providers are part of a team. Additionally, organizations should display signals that the training is supported at the organizational level (Sims et al., 2005).

During Team Training

Once the conditions have been developed to support the implementation of team training and EMS providers are motivated to improve their teamwork, the next step is to implement the chosen team training initiatives in a manner that promotes and enhances learning (Salas et al., 2012; Ubeda-Garcia et al., 2013).

Choose an Instructional Strategy and Training Method

The instructional strategy (i.e., the tools, content) and the training method (e.g., lecture, simulation) chosen to impart team competencies on EMS providers are of paramount importance (Ubeda-Garcia et al., 2013). First, the content chosen should be driven by the need analysis. Specifically, the training content must be designed to cover topics related to the adoption of attitudes, behaviors, and cognitions that underlie effective teamwork in disaster management situations. However, due to the infrequent occurrence and the diversity of disaster events, relying only on the overarching seven Cs of teamwork (Salas et al., 2014) as the sole content of the team training is not sufficient. Rigorous study of the types of specific teamwork competencies that should be trained should be studied in the context of disaster response. Thus, the aforementioned need analysis is of particular importance.

Similarly, when choosing an instructional strategy, defined as the overarching curriculum which combines the content and training methods into an overall intervention, it is necessary to reference the need analysis. Generally, team training can adopt either an information-, demonstration- or a practice-based approach (Salas & Cannon-Bowers, 2001). While the science shows that information-based and demonstration-based learning methods are the most practical to implement, cost efficient, and help to provide a baseline level of knowledge, research shows that these methods have a minimal impact on actual behavior or attitude change (Weaver et al., 2014). Moreover, due to the nature of disaster situations, more advanced learning strategies, such as simulation-based training and on-the-job active learning strategies, including peer-based mentoring and facilitator coaching, are recommended. Practice opportunities, such as simulation, role play, and guided practice, have been found to be most effective in ensuring the transfer of trained competencies to the job (Gregory et al., 2013). Nonetheless, for practice to be maximally effective, it should be structured and scaffolded, incorporate variability, and allow for self-monitoring (Weaver et al., 2010b).

Use Simulation-Based Team Training

The adoption of simulation-based team training (SBTT) is one mechanism by which a diverse set of disasters can be scripted for teamwork practice (Owen et al., 2006; Weaver et al., 2010b). SBTT is an instructional paradigm in which learners can apply targeted team-based KSAs in a controlled simulated environment (Gaba et al., 2001; Salas et al., 2008a). SBTT helps to reinforce both clinical and teamwork knowledge, skills, and attitudes (Rosen et al., 2008) underlying effective teamwork (e.g., closed-loop communication, situational awareness, backup behaviors, shared mental models; Burke et al., 2004; Salas et al., 2008a). SBTT is most effective and well received when it incorporates multiple learning modalities (i.e., information, demonstration, and practice). It has been shown to be effective in both aviation and military settings (Prince & Salas, 1993; Oser et al., 1999; Salas et al., 2006) and has been used in recent years to train healthcare providers (McGaghie et al., 2010).

EMS providers rarely receive any simulation-based training, unlike their emergency medicine physician counterparts (Stoller et al., 2004; DeVita et al., 2005). However, SBTT in these contexts has been shown to improve teamwork as well as clinical performance on the job. As such, EMS should follow suit and begin to implement such event-based SBTT scenarios to more fully prepare EMS personnel for the types of teamwork skills that they may need to enact in a disaster situation.

An effective SBTT scenario targets only a small subset of the desired teamwork training objectives, so as not to overwhelm the participant. To ensure that trainees practice and demonstrate the specific KSAs of interest, a scenario must contain triggers designed to elicit a given teamwork response (see Rosen et al., 2008). These triggers prompt trainees to demonstrate specific team behaviors and allow them multiple opportunities to practice those behaviors, so that they can be evaluated more accurately and have enough opportunities to master the learning objectives. Furthermore, practice opportunities do not need to be high on physical fidelity. In other words, it is not necessary for the environment within which a practice opportunity takes place to look or sound like a real EMS disaster management scenario. Rather, what is more important is psychological fidelity, defined as the extent to which the training prompts psychological processes that are used on the job (Kozlowski & DeShon, 2004).

Simulation scenarios within SBTT must be designed to reflect the diversity of the types of cases that EMS providers encounter in disaster situations. They must integrate commonly encountered scenarios with high-acuity/low-frequency cases (e.g., pediatric gunshot wound). Additionally, scenarios should incorporate cases with both fatal (e.g., resuscitation) and trivial outcomes (e.g., reflux esophagitis) so that teams can learn how to enact team competencies in a diverse set of situations. An event-based approach to stress exposure training (Driskell et al., 2008) is another promising approach to desensitizing EMS to the inherent stressors characteristic of disaster situations. Correspondingly, team self-correction training, which builds team adaptability into practice so that teams can reflect upon and modify strategies to align with contextual and situational factors as they evolve, is another promising avenue to prepare EMS personnel for disaster management situations (West, 2004). When implementing SBTT, it is critical that the SBTT scenarios are developed to match existing clinical skill levels so to ensure that each trainee is challenged to an appropriate degree and can focus on practicing teamwork behaviors. Moreover, integrating a variety of scenarios will provide learners with ample opportunities to exhibit the desired teamwork skills. Feedback to trainees is another element that must be considered. A systematic review of high-fidelity simulation literature identified feedback (including debriefing) as the most important feature of simulation-based medical education (Issenberg et al., 2005). Recent meta-analytic synthesis indicates that team debriefs can improve performance in simulation or on the job by as much as 20–25% (Tannenbaum & Cerasoli, 2013). SBTT is a means for initiating discussions and reflecting on performance, but debriefing activities also need to be conducted after challenging scenarios to reinforce key concepts. Following SBTT, immediate feedback, including specific positive and negative examples, should be provided. Research suggests that the immediate provision of feedback about what areas can be improved and how these improvements can be made positions trainees to better understand the trained competencies and be more likely to successfully enact them on the job (Burke & Hutchins, 2008). That is because feedback informs the team about how they are performing and identifies strengths and weaknesses. To be most effective, feedback must be diagnostic, accurate, specific, and developmental and be provided immediately following scenario participation in a structured debriefing session (Salas et al., 2008b). A facilitator (typically the trainer) should use a diagnostic scenario-based tool to direct debriefings; however, all team members should actively participate and contribute to the identification of learning opportunities and suggestions for remediation.

To ensure a successful debriefing process and learning experience, the facilitator must provide a supportive and open environment (Smith-Jentsch et al., 1998). Debriefing sessions that are perceived as safe and supportive learning environments will elicit open and honest participation from trainees, where learning is the shared objective, rather than criticism or placing blame (Pearson & Smith, 1986; Smith-Jentsch et al., 1998). Team members should focus on a few critical performance issues during the debriefing process: aspects of team performance in the scenario that went well and the factors that contributed to desired outcomes. This allows the team to focus on repeating these successful behaviors in future performance sessions. Remediation opportunities may involve the repetition of a previously performed scenario or a new scenario requiring a similar set of teamwork skills. The credibility of feedback may also be enhanced by using video recordings of performance episodes and using objective checklists. In order to maximize the benefits of debriefings, facilitators should be trained and utilize evidence-based methodology.

AFTER TEAM TRAINING

When training is complete, there is still work to be done to ensure continued success of the team training program. We will now highlight the tools available to evaluate, reinforce, and sustain team training in EMS.

Evaluate

A training program's success cannot be demonstrated if it is not measured and evaluated. Kirkpatrick (1959) recommends evaluating a training program on four levels: reactions to the training (e.g., the extent to which trainees liked the training and/or thought that it was useful), learning that occurred as a result of the training (e.g., changes in knowledge about teamwork competencies), competencies that are transferred to the job (e.g., conflict management skills used during a disaster management situation), and results that occur after the training (e.g., quicker triage times, less mortality). Several observational checklists (e.g., Observational Teamwork Assessment of Surgery [Healey et al., 2004] and the Communication and Teamwork Skills Assessment [Frankel et al., 2007]) and frameworks (e.g., the simulation module for assessment of resident targeted event responses [Rosen et al., 2008]) are available to assist in measuring skill-based learning or teamwork behaviors on the job. However, with the unique conditions of a disaster management context, it is possible that these tools may not be appropriate. Therefore, we recommend development or adaptation of such measurement tools to account for the specific needs of the disaster management context.

Reinforce

Perceptions of utility are critical to enhance trainee motivation and consequently transfer (Colquitt et al., 2000), yet continual reinforcement of the use of teamwork is essential for its continued use (Grossman & Salas, 2011). Reinforcing training helps to ensure that the time spent on training is not wasted. Practical tools can facilitate retention and transfer of the competencies learned in training. However, these tools must be designed with the constraints of a disaster management situation in mind. That is, they should be portable, small, yet accessible. Thus, tools such as pocket guides or brief teamwork reminders printed on the back of providers' identification badges may be optimal choices. Additionally, reinforcement strategies such as the use of regular feedback to trainees on the use or misuse of teamwork (Ford & Weissbein, 1997; Hackman, 1987) and provision of materials and resources such as reminders, policies, and rewards (Grossman & Salas, 2011) are all critical to send the message that the use of teamwork is an essential part of the job. Similarly, supervisors should be encouraged to show support for the training as supervisor support has been found to increase training transfer (Smith-Jentsch et al., 2001). Supervisors should provide incentives for use of the trained competencies on the job. For example, a supervisor may choose to publically reward an employee who displayed good teamwork on the job. This is likely to cue providers to display similar behaviors in future situations, including disaster management situations.

Sustain

Lastly, team training should be sustained so that trained competencies are used continually over time. There is much to learn on best practices for cultivating a long-term culture of teamwork (King & Harden, 2013), yet the evidence from the science of training and organizational behavior states that a cultural climate (i.e., the employee perceptions of organizational norms and values) that is receptive to the training intervention will increase the likelihood that training will be used on the job (Blume et al., 2010). Thereby, it is critical to make the adoption of teamwork and team training along with a supportive culture of the team development intervention a priority. Sustainment can be facilitated by engaging in a number of practices. First, the organization and supervisors must continue to show support for teamwork and use of the trained competencies on the job (Salas et al., 2009). This entails dedicating resources to and committing to use the training. Supervisors—such as station leaders—serve as a model to their subordinates; as such, it is essential that they model use of the trained teamwork competencies so as to show that the competencies are valued over time.

Additionally, refresher training should be conducted as needed. This provides an opportunity for employees to relearn anything that has been forgotten with time. This point is especially important when conducting teamwork training for disaster management situations, which tend to occur infrequently. Furthermore, conducting refresher training helps to ensure that new employees who joined the organization after the initial team training session(s) are also exposed to teamwork training. When considering this effort, it is important to weigh the decrement in trained competencies over time against the cost and required resources of these additional sessions. Meta-analytic literature on skill decay (i.e., loss of retention of learned material or skills over time) shows that learners tend to have significant and large decreases in retention in shorter than 2 weeks' time (Arthur et al., 1998). However, it is likely not feasible, practical, or advisable to retrain employees every 2 weeks. We recommend that agencies set a realistic schedule for refresher training, while also keeping in mind that decrements in retention tend to occur quickly.

CONCLUSION

EMS providers are critical when it comes to managing disasters, saving lives, and providing relief and aid to the injured. It is imperative that EMS professionals are prepared not only as a system but as a team and as individuals. The science of teams has been widely implemented in the healthcare industry, but there is a great need to better understand teamwork in the EMS context. In this chapter, we have reviewed what is known regarding the role of teamwork in EMS and the conditions of the EMS setting that impact teamwork. Additionally, we discuss a way forward for improving teamwork in EMS disaster management contexts via team training. Nonetheless, there is much left to uncover in terms of how to promote the importance of teamwork, how to best implement team training, and how to sustain teamwork in uncertain, rapidly evolving, and tumultuous environments that often characterize the disasters that EMS are called to dispel. We urge researchers to continue to investigate the mechanisms through which teamwork in EMS can be improved, reinforced, maintained, evaluated, and ultimately institutionalized.

ACKNOWLEDGMENTS

The work presented here was supported by funding from the Florida Department of Health Bureau of Emergency Medical Services (grant no. 64018247).

REFERENCES

Armenakis, A., Harris, S. & Feild, H. (2000). Making change permanent: A model for institutionalizing change interventions. *Research in Organizational Change and Development, 12*, 97–128.
Arthur, W., Bennett Jr, W., Stanush, P.L. & McNelly, T.L. (1998). Factors that influence skill decay and retention: A quantitative review and analysis. *Human Performance, 11*, 57–101.
Aubé, C. & Rousseau, V. (2005). Team goal commitment and team effectiveness: The role of task interdependence and supportive behaviors. *Group Dynamics: Theory, Research, and Practice, 9*, 189–204.
Auf der Heide, E. (2006). The importance of disaster management planning. *Annals of Emergency Medicine, 47*, 34–49.
Axtell, C.M., Maitlis, S. & Yearta, S.K. (1997). Predicting immediate and longer-term transfer of training. *Personnel Review, 26*, 201–213.
Barling, J., Christie, A. & Hoption, C. (2011). Leadership. In S. Zedeck (Ed.), *Handbook of industrial and organizational psychology* (Vol. 1, pp. 183–240). Washington, DC: American Psychological Association.

Bayley, K.B., Savitz, L.A., Maddalone, T., Stoner, S.E., Hunt, J.S. & Wells, R. (2007). Evaluation of patient care interventions and recommendations by a transitional care pharmacist. *Therapeutics and Clinical Risk Management*, 3, 695–703.

Beaton, R.D. & Murphy, S.A. (1993). Sources of occupational stress among firefighter/EMTs and firefighter/paramedics and correlations with job-related outcomes. *Prehospital and Disaster Medicine*, 8, 140–150.

Beaton, R.D., Murphy, S.A., Pike, K.C. & Corneil, W. (1997). Social support and network conflict in firefighters and paramedics. *Western Journal of Nursing Research*, 19, 297–313.

Bentley, M.A., Crawford, J.M., Wilkins, J.R., Fernandez, A.R. & Studnek, J.R. (2013). An assessment of depression, anxiety, and stress among nationally certified EMS professionals. *Prehospital Emergency Care*, 17(3), 330–338.

Bigham, B.L., Buick, J.E., Brooks, S.C., Morrison, M., Shojania, K.G. & Morrison, L.J. (2011). Patient safety in emergency medical services: A systematic review of the literature. *Prehospital Emergency Care*, 16, 20–35.

Blume, B.D., Ford, J.K., Baldwin, T.T. & Huang, J.L. (2010). Transfer of training: A meta-analytic review. *Journal of Management*, 36, 1065–1105.

Boksem, M.A., Meijman, T.F. & Lorist, M.M. (2006). Mental fatigue, motivation and action monitoring. *Biological Psychology*, 72, 123–132.

Bowers, C.A., Jentsch, F., Salas, E. & Braun, C.C. (1998). Analyzing communication sequences for team training needs assessment. *Human Factors*, 40, 672–679.

Brown, W.E., Margolis, G. & Levine, R. (2005). Peer evaluation of the professional behaviors of emergency medical technicians. *Prehospital and Disaster Medicine*, 20, 107–114.

Buljac-Samardzic, M., Dekker-van Doorn, C.M., van Wijngaarden, J.D. & van Wijk, K.P. (2010). Interventions to improve team effectiveness: A systematic review. *Health Policy*, 94, 183–195.

Burke, C.S., Salas, E., Wilson-Donnelly, K. & Priest, H. (2004). How to turn a team of experts into an expert medical team: Guidance from the aviation and military communities. *Quality and Safety in Health Care*, 13(suppl 1), i96–i104.

Burke, C.S., Stagl, K.C., Klein, C., Goodwin, G.F., Salas, E. & Halpin, S.M. (2006). What type of leadership behaviors are functional in teams?: A meta-analysis. *The Leadership Quarterly*, 17, 288–307.

Burke, L.A. & Hutchins, H.M. (2008). A study of best practices in training transfer and proposed model of transfer. *Human resource development quarterly*, 19(2), 107–128.

Capella, J., Smith, S., Philp, A., Putnam, T., Gilbert, C., Fry, W. et al. (2010). Teamwork training improves the clinical care of trauma patients. *Journal of Surgical Education*, 67, 439–443.

Cardenas, F.S. & Daccord, Y. (2013). *Ambulance and pre-hospital services in risk situations.* Expert workshop jointly organized by the ICRC and the Mexican Red Cross and held in Toluca, Mexico, from May 20–24, 2013. Oslo, Norway: Norwegian Red Cross.

Carson, J.B., Tesluk, P.E. & Marrone, J.A. (2007). Shared leadership in teams: An investigation of antecedent conditions and performance. *Academy of Management Journal*, 50, 1217–1234.

Colquitt, J.A., LePine, J.A. & Noe, R.A. (2000). Toward an integrative theory of training motivation: A meta-analytic path analysis of 20 years of research. *Journal of Applied Psychology*, 85, 678–707.

Colquitt, J.A., Scott, B.A. & LePine, J.A. (2007). Trust, trustworthiness, and trust propensity: A meta-analytic test of their unique relationships with risk taking and job performance. *Journal of Applied Psychology*, 92, 909–927.

De Bruycker, M., Greco D., Annino I., Stazi, M.A., de Ruggiero, N., Triassi, M., de Kettenis, Y.P. & LeChat, M.F. (1983). The 1980 earthquake in southern Italy: Rescue of trapped victims and mortality. *Bulletin of the World Health Organization*, 61, 1021–1025.

De Bruycker, M., Greco, D., Lechat, M.F., Annino, I., De Ruggiero, N. & Triassi, M. (1985). The 1980 earthquake in southern Italy: Morbidity and mortality. *International Journal of Epidemiology, 14*, 113–117.

DeChurch, L.A. & Mesmer-Magnus, J.R. (2010). The cognitive underpinnings of effective teamwork: A meta-analysis. *Journal of Applied Psychology, 95*, 32–53.

De Dreu, C.K. & Weingart, L.R. (2003). Task versus relationship conflict, team performance, and team member satisfaction: A meta-analysis. *Journal of Applied Psychology, 88*, 741–749.

DeRue, D.S., Nahrgang, J.D., Wellman, N.E.D. & Humphrey, S.E. (2011). Trait and behavioral theories of leadership: An integration and meta-analytic test of their relative validity. *Personnel Psychology, 64*, 7–52.

DeVita, M.A., Schaefer, J., Wang, H. & Dongilli, T. (2005). Improving medical emergency team (MET) performance using a novel curriculum and a computerized human patient simulator. *Quality and Safety in Health Care, 14*, 326–331.

Driskell, J.E., Salas, E., Johnston, J.H. & Wollert, T.N. (2008). Stress exposure training: An event-based approach. *Performance Under Stress*, 271–286.

Edmondson, A.C. (1999). Psychological safety and learning behavior in work teams. *Administrative Science Quarterly, 44*, 350–383.

Edmondson, A.C. (2003). Speaking up in the operating room: How team leaders promote learning in interdisciplinary action teams. *Journal of Management Studies, 40*, 1419–1452.

Ellingson, L.L. (2002). Communication, collaboration, and teamwork among health care professionals. *Communication Research Trends, 21*, 3–21.

Eschmann, N.M., Pirrallo, R.G., Aufderheide, T.P. & Lerner, E.B. (2010). The association between emergency medical services staffing patterns and out-of-hospital cardiac arrest survival. *Prehospital Emergency Care, 14*, 71–77.

Fairbanks, R.J., Crittenden, C.N., O'Gara, K.G., Wilson, M.A., Pennington, E.C., Chin, N.P. & Shah, M.N. (2008). Emergency medical services provider perceptions of the nature of adverse events and near misses in out-of-hospital care: An ethnographic view. *Academic Emergency Medicine, 15*(7), 633–640.

Flin, R. & Maran, N. (2004). Identifying and training non-technical skills for teams in acute medicine. *Quality and Safety in Health Care, 13*, 180–184.

Ford, J.K. & Weissbein, D.A. (1997). Transfer of training: An updated review and analysis. *Performance Improvement Quarterly, 10*, 22–41.

Frankel, A., Gardner, R., Maynard, L. & Kelly, A. (2007). Using the communication and teamwork skills (CATS) assessment to measure health care team performance. *Joint Commission Journal on Quality and Patient Safety, 33*, 549–558.

Gaba, D.M., Howard, S.K., Fish, K.J., Smith, B.E. & Sowb, Y.A. (2001). Simulation-based training in anesthesia crisis resource management (ACRM): A decade of experience. *Simulation & Gaming, 32*, 175–193.

Galbraith, J.R. (1973). Designing complex organizations. Addison-Wesley Longman Publishing Co., Inc.

Goldstein, I.L. & Gilliam, P. (1990). Training system issues in the year 2000. *American Psychologist, 45*(2), 134.

Goldstein, I.L. (1991). Training in work organizations. In M.D. Dunnette, L.M. Hough (Eds.), *Handbook of industrial and organizational psychology* Vol. 2. Palo Alto, CA: Consulting Psychologists Press.

Gregory, M.E., Feitosa, J., Driskell, T., Salas, E. & Vessey, W.B. (2013). Designing, delivering, and evaluating team training in organizations: Principles that work. In E. Salas, S.I. Tannenbaum, D. Cohen & G. Latham (Eds.), *Developing and enhancing teamwork in organizations: Evidence-based best practices and guidelines* (pp. 441–487). San Francisco: Jossey-Bass.

Gregory, M.E., Hughes, A.M., Sonesh, S., Benishek, L.E., Joseph, D.L., King, H.B. & Salas, E. (2014, May). *Team training in healthcare: A meta-analysis and integration.* Poster presented at the 29th Annual Conference for the Society for Industrial and Organizational Psychology, Honolulu, HI.

Grossman, R. & Salas, E. (2011). The transfer of training: What really matters. *International Journal of Training and Development, 15,* 103–120.

Hackman, J.R. (1987). The design of work teams. In J.W. Lorsch (Ed.), *Handbook of organizational behavior* (pp. 315–342). Englewood Cliffs, NJ: Prentice Hall.

Harden, R.M. (2006). Trends and the future of postgraduate medical education. *Emergency Medicine Journal, 23,* 798–802.

Healey, A.N., Undre, S. & Vincent, C.A. (2004). Developing observational measures of performance in surgical teams. *Quality and Safety in Health Care, 13,* i33–i40.

Hollenbeck, J.R., Ilgen, D.R., Sego, D.J., Hedlund, J., Major, D.A. & Phillips, J. (1995). Multilevel theory of team decision making: Decision performance in teams incorporating distributed expertise. *Journal of Applied Psychology, 80,* 292–316.

Hughes, A.M., Gregory, M.E., Joseph, D.L., Sonesh, S.C., Marlow, S.L., Lacerenza, C.N., Benishek, L.E., King, H.B. & Salas, E. (2016). Saving lives: A meta-analysis of team training in healthcare. *Journal of Applied Psychology, 101*(6), 1266–1304.

Hughes, A.M., Sonesh, S., Zajac, S. & Salas, E. (2013). Leveraging HFACS to understand medication error in emergency medical services (EMS): A systematic review. *Proceedings of the Human Factors and Ergonomics Society Annual Meeting, 57,* 1688–1692.

Hughes, A.M., Patterson, P.D., Weaver, M., Gregory, M., Sonesh, S.C., Landsittel, D., Krackhardt, D., Hostler, D., Lazzara, E.H., Vena, J.E., Salas, E. & Yealy, D. (in press). Teammate familiarity, teamwork, and risk of workplace injury in EMS teams. *Journal of Emergency Nursing.*

Issenberg, S.B., McGaghie, W.C., Petrusa, E.R., Lee Gordon, D. & Scalese, R.J. (2005). Features and uses of high-fidelity medical simulations that lead to effective learning: A BEME systematic review. *Medical Teacher, 27,* 10–28.

Jehn, K. (1994). Enhancing effectiveness: An investigation of advantages and disadvantages of value-based intragroup conflict. *International Journal of Conflict Management, 5,* 223–238.

Jehn, K. (1995). A multimethod examination of the benefits and detriments of intragroup conflict. *Administrative Science Quarterly, 40,* 256–282.

Jehn, K. (1997). Affective and cognitive conflict in work groups: Increasing performance through value-based intragroup conflict. In C.K.W. De Dreu & E. Van de Vliert (Eds.), *Using conflict in organizations* (pp. 87–100). London: Sage.

Joint Commission. (2014). Medical device alarm safety in hospitals. *The Joint Commission Sentinel Event Alert, 50,* 1–3.

King, H.B. & Harden, S.W. (2013). Sustainment of teamwork. In E. Salas, K. Frush, D.P. Baker, J.B. Battles, H.B. King & R.L. Wears (Eds.), *Improved patient safety through teamwork and team training* (pp. 188–201). New York: Oxford University Press.

Kirkpatrick, D. (1959). Techniques for evaluating training programs. *Journal of the American Society for Training and Development, 13,* 3–9.

Kotter, J.P. (1995). Leading change: Why transformation efforts fail. *Harvard Business Review, 73,* 59–67.

Kozlowski, S.W.J. & DeShon, R.P. (2004). A psychological fidelity approach to simulation-based training: Theory, research and principles. In S.G. Schiflett, L.R. Elliott, E. Salas & M.D. Coovert (Eds.), *Scaled worlds: Development, validation and applications* (pp. 75–99). Burlington, VT: Ashgate Publishing.

Kozlowski, S.W., Watola, D.J., Jensen, J.M., Kim, B.H. & Botero, I.C. (2009). Developing adaptive teams: A theory of dynamic team leadership. In E. Salas, G.F. Goodwin & C.S. Burke (Eds.), *Team effectiveness in complex organizations: Cross-disciplinary perspectives and approaches* (pp. 113–155). New York: Taylor & Francis.

Lammers, R., Byrwa, M. & Fales, W. (2012). Root causes of errors in a simulated prehospital pediatric emergency. *Academic Emergency Medicine*, *19*, 37–47.

Larson, J.R., Foster-Fishman, P.G. & Keys, C.B. (1994). Discussion of shared and unshared information in decision-making groups. *Journal of Personality and Social Psychology*, *67*, 446–461.

Leonard, M., Graham, S. & Bonacum, D. (2004). The human factor: The critical importance of effective teamwork and communication in providing safe care. *Quality and Safety in Health Care*, *13*, i85–i90.

Marks, M.A., Mathieu, J.E. & Zaccaro, S.J. (2001). A temporally based framework and taxonomy of team processes. *Academy of Management Review*, *26*, 356–376.

McCulloch, P., Mishra, A., Handa, A., Dale, T., Hirst, G. & Catchpole, K. (2009). The effects of aviation-style non-technical skills training on technical performance and outcome in the operating theatre. *Journal of Quality and Safety in Health Care*, *18*, 109–115.

McGaghie, W.C., Issenberg, S.B., Petrusa, E.R. & Scalese, R.J. (2010). A critical review of simulation-based medical education research: 2003–2009. *Medical Education*, *44*, 50–63.

McGrath, J.E. (1984). Studying groups at work: Ten critical needs for theory and practice. In P. Goodman (Ed.), *Designing effective work groups* (pp. 362–391). San Francisco: Jossey-Bass.

Mell, H. & Sztajnkrycer, M. (2004). EMS response to Columbine: Lessons learned. *The Internet Journal of Rescue and Disaster Medicine*, *5*. Retrieved from https://ispub.com /IJRDM/5/1/12573.

Meshkati, N. & Hancock, P.A. (Eds.). (2011). *Human mental workload* (Vol. 52). Elsevier.

Mesmer-Magnus, J.R. & DeChurch, L.A. (2009). Information sharing and team performance: A meta-analysis. *Journal of Applied Psychology*, *94*, 535–546.

Morgeson, F.P., DeRue, D.S. & Karam, E.P. (2010). Leadership in teams: A functional approach to understanding leadership structures and processes. *Journal of Management*, *36*, 5–39.

Neily, J., Mills, P.D., Young-Xu, Y., Carney, B.T., West, P., Berger, D.H. et al. (2010). Association between implementation of a medical team training program and surgical mortality. *Journal of the American Medical Association*, *304*, 1693–1700.

Nguyen, H. & Nguyen, B.D.T. (2008). Portable patient devices, systems, and methods for providing patient aid and preventing medical errors, for monitoring patient use of ingestible medications, and for preventing distribution of counterfeit drugs. *U.S. Patent Application 12/079,080.*

Noakes, T.D., St. Clair Gibson, A. & Lambert, E.V. (2005). From catastrophe to complexity: A novel model of integrative central neural regulation of effort and fatigue during exercise in humans: Summary and conclusions. *British Journal of Sports Medicine*, *124*, 39120–39124.

O'Dea, A., O'Connor, P. & Keogh, I. (2014). A meta-analysis of the effectiveness of crew resource management training in acute care domains. *Postgraduate Medical Journal*, *90*(1070), 699–708.

Oriol, M.D. (2006). Crew resource management: Applications in healthcare organizations. *Journal of Nursing Administration*, *36*, 402–406.

Oser, R., Cannon-Bowers, J.A., Salas, E. & Dwyer, D. (1999) Enhancing human performance in technology-rich environments: Guidelines for scenario-based training. In E. Salas (Ed.), *Human/technology interaction in complex systems* (Vol. 9, pp. 175–202). Greenwich, CT: JAI Press.

Owen, H., Mugford, B., Follows, V. & Pullmer, J. (2006) Comparison of three simulation-based training methods for management of medical emergencies. *Resuscitation*, *71*, 204–211.

Patterson, D.P., Anderson, M.S., Zionts, N.D. & Paris, P.M. (2013). The emergency medical services safety champions. *American Journal of Medical Quality, 28*, 286–291.

Patterson, D.P., Arnold, R.M., Abebe, K., Lave, J.R., Krackhardt, D., Carr, M. et al. (2011a). Variation in emergency medical technician partner familiarity. *Health Services Research, 46*, 1319–1331.

Patterson, P.D., Suffoletto, B.P., Kupas, D.F., Weaver, M.D. & Hostler, D. (2010). Sleep quality and fatigue among prehospital providers. *Prehospital Emergency Care, 14*(2), 187–193.

Patterson, D.P., Weaver, M.D. & Rac, F. (2012). Association between poor sleep, fatigue, and safety outcomes in emergency medical service providers. *Prehospital Emergency Care, 16*, 86–97.

Patterson, P.D., Weaver, M.D., Weaver, S.J., Rosen, M.A., Todorova, G., Weingart, L.R. et al. (2011b). Measuring teamwork and conflict among emergency medical technician personnel. *Prehospital Emergency Care, 16*, 98–108.

Pearce, C.L., Manz, C.C. & Sims, H.P. (2009). Where do we go from here?: Is shared leadership the key to team success? *Organizational Dynamics, 38*, 234–238.

Pearson, M. & Smith, D. (1986). Debriefing in experience-based learning. *Simulation/Games for Learning, 16*,155–172.

Peleg, K. & Pliskin, J.S. (2004). A geographic information system simulation model of EMS: Reducing ambulance response time. *The American Journal of Emergency Medicine, 22*(3), 164–170.

Pianezza, P. (2010). Leadership training. Retrieved from http://www.emsworld.com/article /10319190/ems-leadership-training.

Prince, C. & Salas, E. (1993). Training and research for teamwork in the military aircrew. In E.L., Wiener, B.G. Kanki & R.L. Helmreich (Eds.), *Cockpit resource management* (pp. 337–366). San Diego, CA: Academic Press.

Quarentelli, E. (1985). What is a Disaster? The Need for Clarification in Definition and Conceptualization in Research. *Disasters and mental health: Selected contemporary perspectives* (pp. 41–73). Washington, DC: US Government Printing Office.

Rico, R., Sánchez-Manzanares, M., Gil, F. & Gibson, C. (2008). Team implicit coordination processes: A team knowledge–based approach. *Academy of Management Review, 33*, 163–184.

Rosen, M.A., Salas, E., Silvestri, S., Wu, T.S. & Lazzara, E.H. (2008). A measurement tool for simulation-based training in emergency medicine: The simulation module for assessment of resident targeted event responses (SMARTER) approach. *Simulation in Healthcare, 3*, 170–179.

Saavedra, R., Earley, P.C. & Van Dyne, L. (1993). Complex interdependence in task-performing groups. *Journal of Applied Psychology, 78*, 61–72.

Salas, E., Almeida, S.A., Salisbury, M., King, H., Lazzara, E.H., Lyons, R., Wilson K.A Almeida P.A & McQuillan, R. (2009). What are the critical success factors for team training in health care?. *The Joint Commission Journal on Quality and Patient Safety, 35*(8), 398–405.

Salas, E., Burke, C.S. & Cannon-Bowers, J.A. (2000). Teamwork: Emerging principles. *International Journal of Management Reviews, 2*, 339–356.

Salas, E. & Cannon-Bowers, J.A. (2001). The science of training: A decade of progress. *Annual Review of Psychology, 52*, 471–499.

Salas, E., Cannon-Bowers, J.A. & Johnston, J.H. (1997). How can you turn a team of experts into an expert team?: Emerging training strategies. In C.E. Zsambok & G. Klein (Eds.), *Naturalistic decision making* (pp. 359–370). Mahwah, NJ: Lawrence Erlbaum Associates.

Salas, E., DiazGranados, D., Klein, C., Burke, C.S., Stagl, K.C., Goodwin, G.F. & Halpin, S.M. (2008a). Does team training improve team performance? A meta-analysis. *Human Factors, 50*, 903–933.

Salas, E., Dickinson, T.L., Converse, S.A. & Tannenbaum, S.I. (1992). Toward an understanding of team performance and training. In R.W. Swezey & E. Salas (Eds.), *Teams: Their training and performance* (pp. 3–29). Norwood, NJ: Ablex.

Salas, E., Klein, C., King, H., Salisbury, M., Augenstein, J.S., Birnbach, D.J. et al. (2008b). Debriefing medical teams: 12 evidence-based best practices and tips. *Joint Commission Journal on Quality and Patient Safety, 34*, 518–527.

Salas, E., Priest, H.A., Wilson, K.A. & Burke, C.S. (2006). Scenario-based training: Improving military mission performance and adaptability. Military life: The psychology of serving in peace and combat. *Operational Stress, 2*, 32–53.

Salas, E., Shuffler, M.L., Thayer, A.L., Bedwell, W.L. & Lazzara, E.H. (2014). Understanding and improving teamwork in organizations: A scientifically based practical guide. *Human Resource Management, 54*, 599–622.

Salas, E., Tannenbaum, S.I., Kraiger, K. & Smith-Jentsch, K.A. (2012). The science of training and development in organizations: What matters in practice. *Psychological science in the public interest, 13*(2), 74–101.

Salas, E., Wilson, K.A., Murphy, C.E., King, H. & Salisbury, M. (2008c). Communicating, coordinating, and cooperating when lives depend on it: Tips for teamwork. *Joint Commission Journal for Quality and Patient Safety, 34*, 333–341.

Schein, E.H. (1985). *Organizational Culture and Leadership*. San Francisco: Jossey-Bass.

Schultz, C.H., Koenig, K.L. & Noji, E.K. (1996). A medical disaster response to reduce immediate mortality after an earthquake. *New England Journal of Medicine, 335*, 438–444.

Shaw, J.D., Zhu, J., Duffy, M.K., Scott, K.L., Shih, H. & Susanto, E. (2011). A contingency model of conflict and team effectiveness. *Journal of Applied Psychology, 96*, 391–400.

Sims, D.E. & Salas, E. (2007). When teams fail in organizations: What creates teamwork breakdowns? In J. Langan-Fox, C.L. Cooper & R. Klimoski (Eds.), *Research companion to the dysfunctional workplace: Management challenges and symptoms* (p. 302). Cheltenham, UK: Edward Elgar.

Sims, D.E., Salas, E. & Burke, C.S. (2005). Promoting effective team performance through training. In S.A. Wheelan (Ed.), *The handbook of group research and practice* (pp. 407–425). Thousand Oaks, CA: Sage.

Smith-Jentsch, K.A., Campbell, G.E., Milanovich, D.M. & Reynolds, A.M. (2001). Measuring teamwork mental models to support training needs assessment, development, and evaluation: Two empirical studies. *Journal of Organizational Behavior, 22*, 179–194.

Smith-Jentsch, K.A., Kraiger, K., Cannon-Bowers, J.A. & Salas, E. (2009). Do familiar teammates request and accept more backup? Transactive memory in air traffic control. *Human Factors: The Journal of the Human Factors and Ergonomics Society, 51*, 181–192.

Smith-Jentsch, K.A., Zeisig, R.L., Acton, B. & McPherson, J.A. (1998). Team dimensional training: A strategy for guided team self-correction. Janis, A. (Ed.); Salas, Eduardo (Ed.). (1998). Making decisions under stress: Implications for individual and team training, (pp. 271–297). Washington, DC, US: American Psychological Association, pp. xxiii, 447. http://dx.doi.org/10.1037/10278-010

Sonesh, S., Gregory, M.E., Hughes, A.M., Lacarenza, C., Marlow, S., Cooper, T. et al. (2013). An empirical examination of medication error in emergency medical systems (EMS): Towards a comprehensive taxonomy. Florida Department of Health Bureau of Emergency Medical Services (FDHBEMS, grant no. 64018247).

Stasser, G. & Titus, W. (1985). Pooling of unshared information in group decision making: Biased information sampling during discussion. *Journal of Personality and Social Psychology, 48*, 1467–1478.

Stewart, G.L. (2006). A meta-analytic review of relationships between team design features and team performance. *Journal of Management, 32*, 29–55.

Stoller, J.K., Rose, M., Lee, R., Dolgan, C. & Hoogwerf, B.J. (2004). Teambuilding and leadership training in an internal medicine residency training program. *Journal of General Internal Medicine, 19,* 692–705.

Tannenbaum, S.I. & Cerasoli, C.P. (2013). Do team and individual debriefs enhance performance? A meta-analysis. *Human Factors, 55,* 231–245.

Tannenbaum, S.I., Salas, E. & Cannon-Bowers, J.A. (1996). Promoting team effectiveness. In M. West (Ed.), *Handbook of work group psychology* (pp. 503–529). New York: Wiley-Blackwell.

Tomek, S. (2008). Combined team training. *EMS World.* Retrieved from http://www.ems world.com/article/10321218/combined-team-training.

Ubeda-García, M., Marco-Lajara, B., Sabater-Sempere, V. & García-Lillo, F. (2013). Does training influence organisational performance? Analysis of the Spanish hotel sector. *European Journal of Training and Development, 37*(4), 380–413.

Ulmer, C., Wolman, D.M. & Johns, M.M. (Eds.). (2009). Resident duty hours: Enhancing sleep, supervision, and safety. National Academies Press.

Vernon, A. (2013, May 8). How to foster teamwork on the scene: Response considerations for mass violence incidents. Retrieved December 15, 2014, from http://www.jems.com /article/major-incidents/how-foster-teamwork-shooting-incidents.

Vilensky, D. & MacDonald, R.D. (2011). Communication errors in dispatch of air medical transport. *Prehospital Emergency Care, 15*(1), 39–43.

Wahr, J.A., Prager, R.L., Abernathy, J.H., Martinez, E.A., Salas, E., Seifert, P.C., Groom, R.C., Spiess, B.D., Searles, B.E., Sundt, T.M., Sanchez, J.A., Shappell, S.A., Culig, M.H., Lazzara, E.H., Fitzgerald, D.C., Thourani, V.H., Eghtesady, P., Ikonomidis, J.S., England, M.R., Sellke. F.W. & Nussmeier, N.A. (2013). Patient safety in the cardiac operating room: Human factors and teamwork a scientific statement from the American Heart Association. *Circulation, 128*(10), 1139–1169.

Weaver, S.J., Dy, S.M. & Rosen, M.A. (2014). Team training in healthcare: A narrative synthesis of the literature. *British Medical Journal of Quality and Safety, 23,* 359–372.

Weaver, S.J., Lyons, R., DiazGranados, D., Rosen, M.A., Salas, E., Oglesby, J. et al. (2010a). The anatomy of health care team training and the state of practice: A critical review. *Academic Medicine, 85,* 1746–1760.

Weaver, S.J., Lyons, R., Lazzara, E.H., Rosen, M.A., DiazGranados, D., Grim, J. et al. (2010b). Simulation-based team training (SBTT) at the sharp end: A qualitative study of SBTT design, implementation, and evaluation in healthcare. *Journal of Emergencies, Trauma, and Shock, 3,* 369–377.

West, M. (2004) *Effective teamwork.* Oxford, UK: Blackwell.

Wickens, C.D. & Hollands, J. (2000). Engineering psychology and human performance (3rd ed.). Saddle Brook, NJ: Prentice-Hall.

Wildman, J.L., Thayer, A.L., Pavlas, D., Salas, E., Stewart, J.E. & Howse, W.R. (2012). Team knowledge research: Emerging trends and critical needs. *Human Factors, 54,* 84–111.

Williams, K.A., Rose, W.D., Simon, R. & the Med Teams Consortium. (1999). Teamwork in emergency medical services. *Air Medical Journal, 18,* 149–153.

Winquist, J.R. & Larson, J.R. (1998). Information pooling: When it impacts group decision making. *Journal of Personality and Social Psychology, 74,* 371–377.

Wittenbaum, G.M. (1998). Information sampling in decision-making groups: The impact of members' task-relevant status. *Small Group Research, 29,* 57–84.

Wittenbaum, G.M. (2000). The bias toward discussing shared information: Why are high status group members immune? *Communication Research, 27,* 379–401.

Wittenbaum, G.M. & Stasser, G. (1996). Management of information in small groups. In J.L. Nye & A.M. Browers (Eds.), *What's social about social cognition* (pp. 967–978). Thousand Oaks, CA: Sage.

Xiao, Y., Hunter, W.A., Mackenzie, C.F., Jefferies, N.J. & Horst, R.L. (1996). Task complexity in emergency medical care and its implications for team coordination. *Human Factors, 38*, 636–645.

14 Resilience Engineering in Prehospital Emergency Medical Services

Shawn Pruchnicki and Sidney Dekker

CONTENTS

INTRODUCTION: PART OF A SYSTEM APPROACH

Resilience has always been a critical property of all human (and most other "live") systems, but its more recent use in the safety literature has brought an old term to a new understanding. As such, this newer usage has the potential to be more insightful and, thus, more useful when trying to understand accident causation. Prior to the relatively new concepts of the *new view* of human error and how organizations and people are resilient to failure, investigators relied on a blame-and-train-type mentality. That is, actors at the sharp end, being the last ones involved, are blamed for the event and are retrained, so they will perform better next time. However, more frequently, the actors at the sharp end are either fired and/or prosecuted, which further represents a strictly old view mentality.

This old perspective on accident causation directs the crosshairs of accountability squarely on that of the last human to make a decision prior to the event in question. Organizations with this view blame those that last acted upon their most recent understanding of the situation that they faced, that is, explanations that place the sole responsibility for the event on the shoulders of the human(s) in the system regardless of the system that they were trying to function within. In the aviation domain, where much of the accident causation literature is centered, "pilot error" is the old view language of choice. Fortunately, this has started to abate from within many global investigative organizations and their causation reports. For those few stubborn organizations that remain, this notion of blame offers a trajectory that if you simply fix the errant human, then everything within your simple, tractable, and well-designed system will be fine and return to normal.

However, more recently, researchers and investigators have come to realize that these systems, be they aviation, shipping, or prehospital medical care, are far from simple tractable systems. Despite our best efforts, and, of course, realized in hindsight, not all of these systems are always well designed and they struggle every day to stay within some prescribed boundary of safe operations. The failures, represented in these industries like many others, are not due to a few bad apples (Dekker, 2014). Instead, they remain deeply conflicted, intractable, and actually quite complex (Dekker, 2014). These safety-sensitive industries are very interdependent and do not collapse because of a single point failure. Instead, "failures occur when multiple contributors—each necessary but only jointly sufficient—combine" (Woods & Hollnagel, 2006). These interdependent systems through both expected and unexpected interactions can create unanticipated side effects and sudden dramatic failures (Woods, 2009; Alderson & Doyle, 2010). The problem is not the humans but, rather, the systems that we design around humans and expect them to safely navigate every day. More importantly, the very same humans that we always depend on to shoulder the blame for the rare system failure actually make these difficult systems successful every day because of our resilient capabilities. Majority of the time, the work appears on the outside to be quite effortless when, actually, the opposite is true. Human adaptability and the ability to recognize when procedures no longer fit the conditions are what keep these operations functional. If we are to be successful in understanding error, then we must understand how people learn and adapt to create safety in a world fraught with gaps, hazards, trade-offs, and multiple goals (Cook, Render & Woods, 2000). Prehospital emergency care is a perfect example of a system that is challenged by both potentially rapidly changing environmental conditions and critically ill patients who are cared for in time-compressed scenarios.

RESILIENCE ENGINEERING

Despite the use of the term *resilience engineering* (RE) in more recent safety literature, it is nonetheless used in conflicting ways with different industries using it in dissimilar contexts. One of the founding fathers of the concept of RE, Professor David Woods, offers us no less than four different ways of how resilience is used across the majority of the current literature that he reviewed (Woods, n.d.). These include (1) resilience as rebound from trauma and return to equilibrium, (2) resilience as a synonym for robustness, (3) resilience as opposed to brittleness, and (4) resilience as network architectures that can adapt to future surprises.

RE emphasizes examples of the positive, whereas the focus is with how systems succeed by adapting their performance to the demands within the environment (Hollnagel, Woods & Leveson, 2006). Not only are these environments more than conditions such as weather but also how work is imagined and trained, in addition to all the resources required to be successful. Therefore, the question for our consideration is, how can prehospital medicine ensure success while also acknowledging that people are making the best choices possible when managing everyday work? And, of course, being mindful that they are doing so with incomplete information and pressures such as time and resource constraints, how can we better support the frontline worker knowing that the environment that they work in will always be challenging and under pressure? An old view mentality is deeply unfulfilling and falls short of this goal. Some answer this question by offering "better procedures" and additional training, but we should take care to not delude ourselves into thinking that we can plan for every situation or train for every event.

Historically, other domains such as aviation and, more recently, healthcare have already made that mistake. This is where resilience should enter the discussion to help us understand how systems fail. The "blame-and-train" mentality is left over from a time when simple human error failures were offered as a meaningful explanation. One such example from the past that has not helped anyone better protect complex sociotechnical systems is the notion of a *root cause*. This explanation of failure is left over from when the world of mechanical engineering was used to try to explain system failures as far back as the 1930s. This concept offered hope that if you just keep going back far enough, you can find the initial deterministic reason for this event that you can now act upon.

More recently, and to the detriment of the investigative community, these ideas of causation were only bolstered further by the "Swiss cheese" model (Reason, 1990). Although it offered us discussions about looking at the whole system and not just the human, it also provided new language to further confuse the issue. The notions of "holes lining up" and linear accident trajectories offered no practical usefulness or operational ability to the investigative community when standing among the rubble. It is left over from a time when it was proposed that performance variability throughout the system, humans included, could explain accident causation and illuminate that one cause that we seek to feel better about understanding the unexpected failure that we are investigating.

The current RE literature teaches us that normal human performance variability, which offered old view investigators an explanation for failure, is actually responsible for why things almost always go right. It is seen as a positive as opposed to old thinking, which frames it as a negative function and one that offers an explanation of failure. After all, it is this variability that makes systems function to begin with—as humans are able to manage varying degrees of complexity in uncertain dynamic events faced every day—and most of the time we do it very, very well (Hollnagel & Woods, 2005). We are adaptive and resilient beings.

This leads us to the real question—if failures in complex dynamic sociotechnical systems cannot really be explained away with pointless labels such as "human error" or physiological states such as fatigue (Pruchnicki, Wu & Belenky, 2011), then how are we to understand why these events occur despite everyone's best efforts? RE introduced us to the concept that when trying to understand why systems fail, we need to understand where the operational boundaries of a given operation are placed. We need the ability to understand at any moment in time during an operation where we are now and where we are going in relation to these boundaries. But this is not enough; just as important is how a system monitors the boundary conditions of the current model for competence (e.g., how strategies are matched to demands) and adjusting or expanding that model to better accommodate changing demands (Woods, 2006). These changing demands on a system, sometimes called perturbations, are quite normal for any operation regardless of the domain. A significant component of RE is the task of monitoring organizational decision-making and the assessment of how close the daily operational responses to system perturbations place the system near the point of collapse. In many cases, this proximity to the boundary is closer than realized by many organizations, including prehospital medicine (Woods, 2005). This concept is further captured in the following:

> Resilience, as a form of adaptive capacity, is a system's potential for adaptive action in the future when information varies, conditions change, or when new kinds of events occur, any of which challenge the viability of previous adaptations, models, plans, or assumptions (Woods & Branlat, 2011).

Before we can further explore the areas that define resilience and effectively explain what we mean by the different ways that the term *resilience* is applied, we must first explore the notion of perturbations. What are the ways that systems are challenged which may push them closer to failure? In other words, what are we trying to maintain resilience against?

Extensive research in numerous safety-sensitive operations shows us that there are three primary themes to consider: (1) the regularity of the threat or predictability, (2) the threat's potential to disrupt the system, and (3) the threat's origin, be it internal or external (Westrum, 2006). Essentially, the predictability of the threat is more related to the frequency of occurrence as opposed to true predictive capacity. In other words, your operation faces a particular challenge with some regular frequency that you are knowledgeable about, but as to the exact timing that an incident will occur remains far more elusive. The potential of the threat to disrupt the system is related to the extent of the perturbation and its persistence and ability to reduce the resilience of the system and utilization of resources. Once resources, such as workforce or equipment, are depleted, the system becomes more brittle and likely to decompensate and collapse. Some of these threats to system-wide disruption are internal (i.e., consumption of designed and preplanned resources) and, as such, can be more dangerous because of their proximity located behind the scene as compared to an external threat. Understanding how systems fail and whether there are discernible patterns in the said failure offers at least two advantages—the ability to learn from, recognize, and thus predict failure and the implementation of mitigation strategies in the form of RE practices.

DECOMPENSATION

A greatly underutilized perspective by investigators working in complex sociotechnological systems is the discovery of how work is successfully completed, that is, a focus on what goes right as opposed to when things go wrong. Across every domain, we continue to spend tremendous amounts of financial and human resources to focus on events only when things go wrong. With far more success stories than failures, these paths to success if studied and reviewed in a postevent fashion will help reveal how realistic your current policies and procedures are to the real-world challenges faced by the frontline workers. This is the operational space where accidents are brewing—the difference between the optimal performance predicated by manuals and how work is actually done. This occurs not because the humans are not in line but, rather, the system challenges that push them into an adaptive mode such that the procedure–practice gap widens far enough to be noticeable. Is this gap widening because procedures are completely foreign to the real world, or are the novel challenges being faced such that the frontline workers are working so hard and adapting in such novel ways during the time-compressed event to prevent the system from failing? Moreover, these exercises will also show the adaptive capacities (resilience) of the system and how system degradation was controlled and success was achieved. Once understood, better planning and resource allocation can be adjusted in light of the types of challenges experienced. Understanding how the different types of perturbations were presented to your operation and overcame is of the utmost importance. When the system is pushed toward a safety boundary (thereby reducing the margin of maneuverability), these incidents are valuable because they provide information about what stretches the system and how well the system can stretch (Woods & Cook, 2006). How well does your organization or more specifically your employees and procedures adapt to these challenges? During this time, when the time for recovery increases and/or the level recovered by the system decreases, a system is exhausting its ability to handle growing or repeated challenges. This is how we study adaptive capacity and react to our system design based on these observations.

Regardless of the cause of a system-wide decompensation and possible failure, there are inherent characteristics to the patterns experienced. Typically, as designed, complex systems frequently experience perturbations (normative) and the designed adaptive capacities and resilience abilities prevent failure. This is what we plan for, as these are challenges that we have seen before and also expect in the future. Examples might include second alarms calling for additional personnel or heavy equipment brought to the scene to manage an extrication that is now more clearly understood to be more difficult than first appreciated. These adaptations that are constructed under time pressure are the hallmark of a system that is designed to operate in a complex dynamic environment. Typically, most operations in the prehospital EMS recognize these challenges and provide both training and procedures to match when recognized by on-scene personnel.

Due to the dynamic nature of complex sociotechnical systems such as EMS prehospital care and the challenges faced in this unique environment, humans are well designed to function within this space. That is, we bring natural adaptive capacity as part of our normal working ability. We are able to sense changing conditions (challenges) in our environment, the patient's condition, or simply our personal safety while on the scene. But despite this capacity, there are times when our capabilities and the procedures that we follow in addition to the very nature of our work are not enough to prevent deterioration of our scene management as we are pushed toward a boundary and failure. So how do complex systems such as EMS fail? How do we go from managing effectively and operating in offensive proactive ways to quickly establishing a defensive posture—leading to an increased amount of work to simply keep up? How does our system collapse when challenged and what do those events look like from within the organization? To understand how these systems collapse, let us examine patterns of how complex systems fail and the patterns that are commonly observed. If we are able to understand these patterns, then they might have a predictive value, which greatly enhances our foresight and might enable us to better poise our operations away from failure as it is developing.

HOW SYSTEMS FAIL

Complex sociotechnical systems such as prehospital EMS are, in many ways, like a living organism. That is, they have a certain degree of robustness and a certain degree of brittleness but, in all cases, make adaptations to new stresses and challenges (perturbations). Some of these adaptations are successful and some are not. For those cases that are successful (increased fitness), these adaptations alter the dynamics between the components of the system that, in effect, restructures the system as a whole. For example, we learn lessons from an event that yields procedural changes in addition to maybe new training. While many of these changes (adaptations) are beneficial in the short term, they may be problematic in the long term as these adaptations once complete may no longer fit new environments and/or new challenges. Essentially, the management of any system such as prehospital EMS entails achieving a balance between brittleness and resilience. However, this balance always occurs in a resource-constrained environment as no agency has unlimited resources regardless of the challenges that they face. Trade-offs have to be made to achieve these balances with both short-term and long-term considerations in mind. The goal is not to be able to handle everything—the goal is to recognize when the system is unable to handle events before they occur.

Some of the original research and work that has helped us understand these trade-offs and how systems are not always able to cope with both acute and chronic challenges were performed in urban firefighting and crisis management (cf. Woods & Cook, 2006). Before organizations can better understand how to make more successful trade-offs in both near- and far-term decision-making, a better understanding of how systems fail is required.

Regardless of the type of system-wide failure that occurs after a system is challenged, they can be usually be understood in hindsight by one of the following categorizations (Woods & Branlat, 2011):

- Because plans and procedures have fundamental limits
- Because the environment changes over time and in surprising ways
- Because the system itself adapts around success given changing pressures and expectations for performance

FUNDAMENTAL LIMITS

In any safety-sensitive domain, including prehospital EMS, procedures and policies play an important role in protecting both us and our patients. This material serves as bedrock for training in both the classroom and the field. However, like aviation and healthcare, the situations that we face on real calls are never the same and, by their very nature, cannot be preplanned in any precise detail with any true level of confidence. Thus, to manage these surprises successfully, we are adaptable and possess the ability to perceive subtle change in our situation, thereby allowing us to adapt our tactical strategies to accommodate. Despite this, system-wide failures can still occur because we rely too heavily on written guidance and their typical inherent limitations restrict us in our adaptable ability.

ENVIRONMENTAL CHANGES

Beyond the procedures we put in place, the characteristics of the environment that we operate in are quite variable and dynamic. The environmental working conditions include not only weather-related considerations but also the social–technical environment in which we are immersed, that is, how people are interacting with each other and the required technologies in place while trying to negotiate the challenges that they face together. Some of these conditions have the potential to change faster than others but will eventually change nonetheless. These alterations can sometimes be predictable, while in other times they can lead to completely surprising outcomes. In any complex system such as prehospital medicine, all variables are unknown; thus, their cumulative effect is unpredictable. The assumptions with which we are operating are no longer valid due to this cumulative effect. Thus, the decisions that we make, which make sense based on previous mental models, become faulty in hindsight. Further failures can occur because the system has changed in ways not foreseen.

SYSTEM-WIDE ADAPTATION

The individuals in any social–technical system are able to adapt to challenges and surprise, but the very nature of the system changes with these choices. That is to say, our adaptive behavior can alter the system and how we are functioning within it. While adaptive behavior can help any system function more effectively in new situations and challenges, it can be maladaptive on others. New pressures and performances that are part of any dynamic environment drive this behavior. All systems are at risk for this type of problem. Although it is healthy to assess your operation's fitness to face new challenges, based on this assessment, we occasionally after an accident make sweeping changes to our operation. However, before implementation, it would be wise to consider what new vulnerabilities are we now exposed to and/or where have we relocated patterns of failure.

SYSTEM COLLAPSE

Now that we understand better how systems fail, let us explore the patterns with which this occurs. Woods and Branlat (2011) describe that there are three basic patterns with which adaptive systems collapse: (1) decompensation, (2) working at cross-purposes, and (3) getting stuck in outdated behaviors.

Decompensation is probably the most commonly understood and realized in prehospital EMS postincident evaluations. This pattern is one where challenges to the system (on-scene operations) grow and cascade faster than expected, thereby eventually overwhelming the adaptive capacities of the organizations involved with the event. This includes not only those on-scene in direct contact with patients but also others in the system, such as dispatch and aeromedical assistance. Within the pattern of decompensation, there are two phases. The first is the appearance of the system and the actors displaying the adaptive capacity that make humans unique. On the surface, it appears to be nothing more than successful adaptation. However, underneath what is actually happening is that the adaptive capacity of the system is effectively hiding the presence and true extent of the disturbance that is challenging the operation. Thus, as resources, expertise, and capability are being used, the true depth and capacity of the disturbance are masked and not fully appreciated. In other words, we adapt so well that the depth of the system stress is not obvious to the actors involved.

As the system marches faster toward the margin of safety and eventual failure, the very tools that humans bring in their adaptive capacity actually hide the true depth of the disturbance prior to failure. The second phase entails the failure process itself once the disturbance overcomes the adaptive capacity because either the resources are exhausted or the tempo becomes too great. This failure pattern typically occurs because the operation falls behind in tempo and/or the organization is unable to recognize that new modes of functioning are required as the event is changing. A requirement for new modes can occur because the adaptations offered by the front-line worker change the event itself or simply the dynamic and unpredictable nature of prehospital medicine in general. An example might include a multiple victim trauma event with worsening weather and delayed second or third alarms. Factors that are more specific might include rapidly deteriorating patients, long transport times, and reduction in available medications and IV fluids. All of these components can, in many cases, be managed, but any operation, no matter how well stocked, has limits and can become overwhelmed and collapse under the right circumstances.

When applying an understanding of decompensation in prehospital medicine, we are referring to the loss of patients that had the potential to be salvaged and/or responder injuries including both physical and psychological. The challenge for those in charge of these operations is to discriminate between adaptive behavior that is successful control and behavior that is a sign of pending failure (Woods & Branlat, 2011). What makes this exceptionally difficult is that this must be monitored both at the individual level and at the level of the entire operation across numerous levels of command structure. For example, fireground leadership must observe both at above their leadership position (i.e., chief or other high-ranking officials that are involved) and throughout the command structure below them for these patterns with the ability to distinguish between adaptive behavior and signs of pending failure.

Examples of failure above their command position would be unanswered requests for more support and below their position might be crews and local leadership (interior attack leaders) becoming more aggressive in their tactical operation. This increase in performance might very well indicate that the event is escalating, at least on a local level, and other signs throughout the operation would need to be considered to determine if the operation is becoming more brittle or these are locally required increases in performance that are needed to render a situation more controllable. Those at the individual level will need to consider which is the case and transmit this increased effort to higher-ranking members so that they can consider if this local effect is, in fact, local or part of a wider level problem with the entire operation.

Professionals such as EMS workers and other type A personalities are poorly calibrated to recognize and ask for help when decompensation patterns emerge at the individual level. This behavior is not unique to EMS but is also seen on the flight deck of commercial aircraft during emergency procedures. When pilots do not communicate that they are becoming overwhelmed, it only serves to further mask the true extent of the disturbance. This not only delays recognition for those in command but also delays when extra resources are summoned and the window of opportunity to prevent the failure narrows. The key to preventing this is to recognize from a command perspective how hard your resources are performing and if their efforts are increasing even if they are unable to recognize it themselves. It is important to understand how insidious this type of failure can be despite awareness such as recognizing what an "all-hands"-type call means to the stress on your operation. Unfortunately, despite this cognizance, signals are still missed as these are very difficult situations to detect in real time as they are not as clear as an all-hands-type call.

WORKING AT CROSS-PURPOSES

Many emergency operations including prehospital care, firefighting, and in-hospital medicine all require a supervisory command structure to maintain not only effective use of resources but also protection of life and property (Bigley & Roberts, 2001; Buck, Trainor & Aguirre, 2006). However, effective and meaningful communication is the linchpin of any incident command structure and working at cross-purposes describes the inability to coordinate between and across different groups in the command structure either upward or downward within the hierarchy. Examples include activities accomplished at lower echelons in the structure that, although very successful locally, are maladaptive and detrimental for higher-level needs and requirements at that moment in the course of a given event. An example might include an interior attack team dealing with a structure fire pressing farther into the structure and who have been initially ordered to be on the scene and is now pushing the fire toward another team located elsewhere in the structure. This second team might have been placed later by the command structure, and they might not have the resources to take on more fire that is now heading their way.

The reverse can occur as well where commanders higher in the organizational structure determine courses of action that fulfill bigger picture needs but could actually be disastrous for the individual frontline worker. An example might be the fireground commander assigning a crew to a lower-priority action when an interior attack team should be backed up due to an exceptional aggressive fire situation. However, it is important to understand that these conflicts do not always flow in a vertical fashion within the command structure but can also occur horizontally. For example, in a firefighting event, ladder companies performing appropriate actions for a situation that they are confronting can at times be quite troublesome for engine companies elsewhere within the structure. And the reverse is true as well. These are cases where one group inadvertently reduces the margin of maneuverability from failure of another group. One EMS example within the same group might include conflicts between extrication efforts and patient stabilization, which frequently occur simultaneously. Other examples across groups might include conflicts between on-scene police with their own set of priorities and EMS/fire with yet another.

All of these problems highlight the challenges of multifaceted disaster operations that are composed of numerous organizations all trying to work in concert to achieve both local and large-scale needs. Despite how well organized communication efforts might be as part of a preplan effort, the challenges found in these dynamic events can still overcome the most well-designed system.

GETTING STUCK IN OUTDATED BEHAVIORS

The third pattern of decompensation is as one where systems or specifically organizations functioning within a system fail to learn. This can occur on both micro (on-scene) and macro (long-term) scales. In this case, patterns of adaptive capacity are used successfully and thus remain in place despite changing conditions, such that other adaptive strategies should now be considered. The primary challenge is that these decisions are made in uncertain environments and unknown outcomes that are discovered only in hindsight. These missed opportunities to replan constitute sources of failure (Woods & Shattuck, 2000). All of these adaptive traps truly reflect the very nature of complex work in resource-constrained environments under time pressures with uncertainty in outcome always present.

RE PRACTICES

Organizational resilience is seen not as a property, but as a capability: A capability to recognize the boundaries of safe operations, and although abnormal operations may stress feedback mechanisms possibly even causing a loss of control, the resilient system will have the capability to steer back from them in a controlled manner before a system-wide collapse may occur (Dekker & Pruchnicki, 2014).

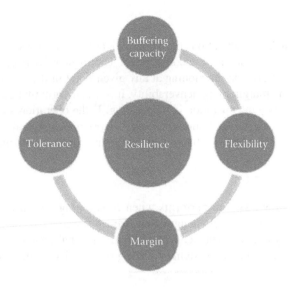

FIGURE 14.1   Properties of resilience. (Adapted from Woods, D.D., *Resilience Engineering: Concepts and Precepts*, Ashgate, Burlington, VT, 2006.)

So what does RE mean from an operational perspective? What are the components that we can focus on either while creating a safer prehospital working environment or when the worst has happened and we are left to literally pick up the pieces? A good starting point could be any of four primary areas of any operation (Figure 14.1) that can offer insight on either side of the accident, that is, either pre or post. To be useful, this requires a thorough examination of the system's buffer capacity, overall flexibility, margin, and tolerance when faced with perturbations to the system as a whole.

BUFFERING CAPACITY

Here, we are examining the size of the known disruptive events that the system can adapt to before signs of decreased performance or a complete collapse of the system, which may include worker or patient harm. Here, we are asking the question, how much can we really handle and where are we the weakest?

FLEXIBILITY

Any system confronted with challenges will try to adapt to these stressors, which may be on both local and more extensive global scales. Regardless of the extent of the disturbance throughout the system, the system's flexibility versus its stiffness when faced with these challenges helps us predict how well it might face other perturbations in the near future. This property tells us more about how well a system will restructure itself when required to do so and how challenging this might be from an organizational perspective.

MARGIN

As discussed earlier, organizations have operational boundaries that define where the system starts to break down or decompensate. The distance between this boundary and where the operation is functioning at any given point in the timeline of the event is described as the margin of maneuverability. It is this margin that describes just how close or precariously the operation is functioning. Is the operation staying effectively away from these performance boundaries and collapse or is it typically operating functioning right up to the point of collapse with little to no wiggle room (maneuverability)?

TOLERANCE

This describes how a system performs when functioning near a boundary and the event in question provides challenges that exceed its adaptive capacity. Does the system gracefully degrade when confronted with this challenge or does it result in a massive collapse of the entire system occurring because its balance is so precarious?

FOUR COMPONENTS OF RE

As discussed at the beginning of the chapter, the usage of the term *resilience* has increased dramatically in the safety literature. However, its usage is rather inconsistent and even at times conflicting as reviewed extensively by Woods (n.d.). In this paper, Professor Woods groups the various uses of resilience around four basic concepts: (1) resilience as rebound from trauma and return to equilibrium, (2) resilience as a synonym for robustness, (3) resilience as opposed to brittleness, and (4) resilience as network architectures that can adapt to future surprises. Let us examine each of these closer.

RESILIENCE AS A REBOUND

When applied as a notion of rebound, resilience helps us understand how organizations, some more than others, recover from a surprise event. Quite possibly the most common description of resilience, it helps us understand not so much how the system actually recovers, the process involved, or the alterations to the overall system state but, rather, the amount of resources available to manage surprise. It is about the volume and type of resources available in addition to the extent of their deployability when faced with surprise (Woods, n.d.). For example, in an EMS environment, these resources can typically be thought of as second alarms, additional companies for workforce, and even requesting for aeromedical support. These examples, and there are many others depending on your specific operation, are ones of additional resources at the ready waiting to be mobilized into action. They are essentially standing by at all times waiting for the request.

RESILIENCE AS ROBUSTNESS

Although seeming to be similar in meaning, a common misconception in the literature is that these two terms are interchangeable when, in fact, they are not. This interchangeable usage in both research and operational settings only serves to worsen this confound. The term should be correctly understood to mean a representation in an increased ability to absorb perturbations (Woods, n.d.), that is, a type of control for cases where the disturbances are well modeled, planned for, and executable. For example, in EMS operations, sending multiple vehicles (such as several ambulances) with trained responders to a scene where multiple victims are known to exist represents a type of robustness to the operation. Having known multiple victims automatically triggers dispatch to use a specific run card, resulting in an appropriate workforce response. To be robust is to be in a position as an organization where an operation is in position enough to handle numerous known possible event types and having a system in place to address them before they occur. This represents known or reasonably suspected types of responses and being prepared with the infrastructure at the ready to manage them at a moment's notice. It is important to understand that EMS operations cannot prepare for every type of possible event, so there are always trade-offs between what are the most likely events that will challenge the system and what internal resources (workforce, financial, hardware, etc.) could be available. Robust control works for only cases where the disturbances are understood and well modeled (Alderson & Doyle, 2010).

A more advanced perspective, and perhaps an alarming one, is that the more robust you make an operation, that is, the more prepared to handle certain types of events that your organization thinks that are possible, the less prepared you might be to handle other types of unrealized events (Alderson & Doyle, 2010; Hoffman & Woods, 2011). Additionally, when speaking of robustness, the focus should be on the system handling expected challenges as they occur. Resilience in this sense does not enter the picture when what is experienced is not understood nor well modeled and the system as a whole moves closer to its safe operational boundaries.

RESILIENCE VERSUS BRITTLENESS

Taken from materials science, the term *brittle* implies that after being exposed to enough stress, any material will eventually break (the system will collapse). When resilience is considered as opposed to brittleness, we are observing how a system can extend its abilities or adaptive capacity in the face of surprise. This usage of the concept is very different from previous discussions where challenges were met with preplanned robustness or when rebounding from an event after additional support was summoned. This perspective asks the question, how does a system function and change when events produce challenges at and beyond its boundaries? (Woods, n.d.). Systems that are functioning in this realm are anticipating bottlenecks in how the challenges are taxing the operation and thus are already learning and acting upon this knowledge to reduce any reduction in the margin of maneuverability from any safety boundary. They are able to adjust to the changing event as it challenges their operation and are ready to adjust their way of thinking about the operation as it experiences this expansion in its capabilities in real time.

An EMS example might be a large structure fire, and realizing that as the fire worsens and more victims are rescued, the system's capabilities are almost exceeded. As such, the on-scene leadership not only calls for more EMS units but also starts alerting hospitals for the pending and unpredictable influx of patients. Additionally, dispatch may also choose to keep an EMS unit (both vehicle and personnel) solely available and strategically located in the run district to address any possible response needs yet not realized on other potential scenes. The ability of the system or rather the lack of it is seen as a potential bottleneck that could risk a firefighter's or patient's life, and a resilient system is already planning for this contingency as the situation worsens and begins to overtax their operation.

Other extensibility outcomes of this example might include calling in off-duty personnel to begin backing up the stations for the routine medical calls that will still challenge the overall system in addition to either placing on standby or launching an aeromedical helicopter to the scene depending on the surrounding environment and distance to the nearest trauma centers. This use of the term describes reaching the limit of a systems capability, and as the personnel are learning from how the system has arrived in this position, they are not only making adjustments to extend the current capability but also starting to anticipate additional originally unforeseen events that could push the system to its breaking point.

RESILIENCE AS ADAPTATION

Resilience in this setting considers the system as a whole and its ability to manage/ regulate adaptive capacities of the various layers within its network to produce sustained adaptability (Woods, n.d.). Important to this perspective is that all systems, regardless of how adaptive they are, are always managing trade-offs. Those that are more successful than others are able to position themselves such that when making these choices, they are always continuing to adjust their resilience so that new challenges will not overwhelm and collapse the operation. For example, an adaptive EMS system would respond to a large-scale event and while doing so would be examining their ability to still handle smaller emergency responses simultaneously. They would be considering where the resource allocation (trade-offs) for the larger event has made them vulnerable for any other responses that might occur. What sets these types of systems apart from others is their ability to recognize while being faced with a massive resource-consuming event, they are still thinking about what they can still handle with either this response or anything along the entire spectrum that could occur any minute. They are not only looking for any potential event that could push them further toward a boundary of safe operation and collapse but also reexamining the current operation to ensure that the trade-offs made on the scene are understood as to their effects on not only the local landscape but also the entire operation and its overall adaptability.

WHY EMS IS DIFFERENT

Prehospital care presents practitioners with challenges that are both unique to healthcare and work in dangerous environments, that is, an amalgamation of the exceptional challenges faced in both workplaces. Nemeth, Cook, and Woods (2004) describe healthcare as "hypercomplex," one that possesses unique complexity and variety and is driven by numerous factors such as unstable patients and a recent explosion in healthcare knowledge/technology. We would offer that this description should be applied to prehospital care as well and maybe for greater reasons. Prehospital emergency care offers almost all of the challenges faced in an emergency room setting but with more layers of unpredictable variety and complexity.

To make matters more challenging, resources in prehospital care are typically stretched even further with the addition of cross scale communication considerations through command hierarchies and on-scene environmental dangers. In many ways, prehospital medicine is as dissimilar to hospital-based emergency medicine as it is similar to not only the specific tasks but also the nuances of how those tasks are completed. On a smaller scale, just the very nature of the work, although still essentially emergency medicine, is very different despite the commonalities.

For example, emergency room workers do not face ultraextreme tasks such as lying in a ditch in the middle of the night while trying to intubate a trauma victim during a thunderstorm. These types of tasks are not rare in prehospital medicine and should force us to consider that these types of prehospital operations are not simply the relocation of the emergency room to the field. Rather this unique system requires that it be viewed from a lens different from any other safety-sensitive operations in healthcare. When emergency rooms are saturated with patients, they are able to redirect their incoming patients to other hospitals in an effort to control their resource allocation for the patients that they are treating, thereby adding to their resilience. However, this type of system regulation that emergency room-based medicine has at their disposal is quite absent in prehospital care. When a call comes in, someone has to respond and they have to do so in a timely manner as opposed to telling the caller to call someone else or call back. Neither of these is an option regardless of the available resources or the level of system-wide resilience—this call has to be managed and resources have to be expended.

Due to these factors faced everyday by prehospital workers, understanding resilience and the degree with which it is present and changes over time is paramount in protecting both our patients and ourselves. Woods and Branlat (2011) describe in detail how challenges in urban firefighting exemplify situations within which tasks and roles are highly distributed and interdependent. These same challenges are realized in prehospital medicine as they expose a working system to the difficulty of maintaining synchronization yet having the flexibility to address ever-changing demands. Unlike hospital-based medicine, which is a static design, prehospital medicine care is evolving during the operation as numerous organizations and companies are all trying to operate in a shared environment that is both dynamic and dangerous. Because of these challenges, one cannot simply think of prehospital care as a relocation of hospital-based emergency medicine.

In this environment, to be considered truly resilient, one must understand the evolving trajectories that push a system toward operational boundaries only to be reversed at a moment's notice, that is, the ability to steer back in a controlled manner while always staying aware of additional potential challenges that will arrive unannounced and further stress the entire operation. This is the landscape where successful systems operate as they always remain alert to the potential of failure and never rest on the success of the past but, rather, the possibility of failure today. Let us never forget that "the past seems incredible, the future implausible" (Woods & Cook, 2002).

RECOMMENDATIONS

RE allows us a novel way to view our systems in both foresight and hindsight, That is, in foresight, we can better understand how systems fail and we can more effectively monitor a system under stress as it marches toward its operational boundary. In hindsight, we abandon the old view mentality and have a more system-wide appreciation of how the human element functions within an operational space that has been determined to have failed. In summary, there are several observations that can help us better understand and apply the concept of RE.

- Systems such as prehospital medicine and many others are deeply conflicted, resource limited, and under extreme time pressures with very real dangers to not only the patients but also the employees involved in administering this care. These systems are not tractable, and the humans involved are not a source of weakness but, rather, a source of adaptability and resilience. In short, it is the humans that make these systems as safe as they are.
- RE has several definitions in the literature but essentially better prepares us to emphasize examples of the positive attributes in our organizations and give us a framework to understand how the actors in the system adapt their performance to match the demands of the environment and the changing landscape of the event. However, understanding how systems fail and the types of patterns that we might be able to recognize in foresight offers important guidance in the overall system-wide observability and its relationship to its boundary of safe operation. Some of these include decompensation, working at cross-purposes, and being stuck in outdated behaviors; all have unique patterns and different ways that resilience can help stave off system collapse.
- In an attempt to operationalize RE, a system's capabilities within the areas of buffering capacity, overall flexibility, and margin and tolerance to perturbations are probed. By examining your operation in depth within this framework, you will be in a better position to understand your strengths and weakness and the scenarios where you are the most vulnerable.

EMS is different from many complex sociotechnical systems, even from that of emergency medicine, which is emulated in some basic components. Healthcare, which has been described as hypercomplex, is never more true than for EMS operations as its unique complexity and variety and ability to rapidly deteriorate in the face of new challenges. RE offers a beacon of light in an otherwise dark understanding of how and why are seemingly well-thought-out systems can spontaneously and catastrophically fail with the sudden loss of human life.

REFERENCES

Alderson, D.L. & Doyle, J.C. (2010). Contrasting views of complexity and their implications for network-centric infrastructures. *IEEE Transactions on Systems, Man and Cybernetics, Part A: Systems and Humans*, 40(4), 839–852.

Bigley, G.A. & Roberts, K.H. (2001). The incident command system: High-reliability organizing for complex and volatile task environments. *Academy of Management Journal*, *44*(6), 1281–1299.

Buck, D.A., Trainor, J.E. & Aguirre, B.E. (2006). A critical evaluation of the incident command system and NIMS. *Journal of Homeland Security and Emergency Management*, *3*(3), 1–27.

Cook, R.I., Render, M. & Woods, D.D. (2000). Gaps in the continuity of care and progress on patient safety. *British Medical Journal*, *320*(7237), 791.

Dekker, S. (2014). *The field guide to understanding human error*. Farnham, Surrey, UK: Ashgate.

Dekker, S. & Pruchnicki, S. (2014). Drifting into failure: Theorising the dynamics of disaster incubation. *Theoretical Issues in Ergonomics Science*, *15*(6), 534–544.

Hoffman, R.R. & Woods, D.D. (2011). Beyond Simon's slice: Five fundamental trade-offs that bound the performance of macrocognitive work systems. *IEEE Intelligent Systems*, *26*(6), 67–71.

Hollnagel, E. & Woods, D.D. (2005). *Joint cognitive systems: Foundations of cognitive systems engineering*. Boca Raton, FL: CRC Press.

Hollnagel, E., Woods, D.D. & Leveson, N. (2006). *Resilience engineering: Concepts and precepts*. Burlington, VT: Ashgate.

Nemeth, C.P., Cook, R.I. & Woods, D.D. (2004). The messy details: Insights from the study of technical work in healthcare. *IEEE Transactions on Systems Man and Cybernetics, Part A: Systems and Humans*, *34*(6), 689–692.

Pruchnicki, S.A., Wu, L.J. & Belenky, G. (2011). An exploration of the utility of mathematical modeling predicting fatigue from sleep/wake history and circadian phase applied in accident analysis and prevention: The crash of Comair Flight 5191. *Accident Analysis & Prevention*, *43*(3), 1056–1061.

Reason, J. (1990). *Human error*. Cambridge, UK: Cambridge University Press.

Westrum, R. (2006). Typology of resilience situations. In E. Hollnagel, D. Woods & N. Leveson (Eds.), *Resilience engineering: Concepts and precepts* (pp. 56–65). Aldershot, UK: Ashgate.

Woods, D. (n.d.) *Four concepts for resilience and their implications for system safety in the face of complexity*. Columbus, OH.

Woods, D.D. (2005). Creating foresight: Lessons for enhancing resilience from Columbia. In M. Farjoun & W.H. Starbuck (Eds.), *Organization at the limit: NASA and the Columbia disaster* (pp. 289–308). Malden, MA: Blackwell.

Woods, D.D. (2006). Essential characteristics of resilience. In E. Hollnagel, D. Woods & N. Leveson (Eds.), *Resilience engineering: Concepts and precepts* (pp. 21–34). Burlington, VT: Ashgate.

Woods, D.D. (2009). Escaping failures of foresight. *Safety Science*, *47*(4), 498–501.

Woods, D.D. & Branlat, M. (2011). Basic patterns in how adaptive systems fail. *Resilience engineering in practice* (pp. 127–144). Aldershot, UK: Ashgate.

Woods, D.D. & Cook, R.I. (2002). Nine steps to move forward from error. *Cognition, Technology & Work*, *4*(2), 137–144.

Woods, D.D. & Cook, R.I. (2006). Incidents—Markers of resilience or brittleness. In E. Hollnagel, D. Woods & N. Leveson (Eds.), *Resilience engineering: Concepts and precepts* (pp. 69–76). Aldershot, UK: Ashgate.

Woods, D.D. & Hollnagel, E. (2006). Resilience engineering concepts. In E. Hollnagel, D. Woods & N. Leveson (Eds.). *Resilience engineering: Concepts and precepts* (pp. 1–16). Aldershot, UK: Ashgate.

Woods, D.D. & Shattuck, L.G. (2000). Distant supervision—Local action given the potential for surprise. *Cognition, Technology & Work*, *2*(4), 242–245.

15 Emergency Medical Service Occupational Safety

Issues, Implications, and Remedies

Brian J. Maguire

CONTENTS

INTRODUCTION

In the United States alone, EMS personnel respond to 31 million emergency medical calls for assistance per year.[1] This makes the workforce a significant component of the US healthcare system, as well as the medical, public health, public safety, and disaster management systems. In addition to their emergency work, EMS personnel are increasingly involved with community health and injury prevention projects.[2-4]

EMS personnel may be paid or unpaid (volunteer) and, in the United States, are generally classified as either (a) EMTs, who are basic life support providers, or (b) paramedics, who are advanced life support providers. There are about a million EMS personnel in the United States; about 80% are volunteers and about 20% are paramedics.[1]

Recent research from Australia has shown that there is no occupational group in the country with a higher injury or fatality rate than EMS personnel.[5] In the United States, EMS personnel have a fatality rate similar to those for police and firefighters[6] and an injury rate much higher than the national average and higher than the rates for police and firefighters.[7]

This chapter will explore EMS occupational safety issues, the implications of those safety issues, and potential remedies.

WHY THIS IS IMPORTANT

Joseph "Neal" Sherman was a 25-year-old paramedic working in Virginia. Neal and his partner had worked together the day before. After their shift, Neal went home and his partner went to the volunteer firehouse where he worked all night. The following day, Neal was taking care of a patient in the back of the ambulance when his partner fell asleep at the wheel. Neal was killed in the crash. He left behind a wife and an unborn child.[8]

Each year in the United States, 10 paramedics are killed and over 5000 are injured.[9] Not only does this have a devastating effect on the workers, their families, friends, and coworkers, it also has an effect on the operational readiness and cost of EMS services. Little is known about the costs of EMS occupational injuries. However, if, as some studies have found, a third of the workforce is injured each year, the problem is extremely costly in terms of real dollars as well as the costs of overtime, insurance, and replacement workers.

Further, the risks are not only to paramedics and EMS agencies but also to our communities. One study found that about 70% of the victims of fatal ambulance crashes were patients, ambulance passengers, passengers of other vehicles, or pedestrians.[6] Maguire found that there were about 80 victims of ambulance crashes per year (fatal and nonfatal).[10] If the same 70% average is true for this group, ambulance crashes are killing and injuring about 60 civilians each year in the United States. These figures do not include potential wake-effect crashes. For example, Clawson et al. found that the number of vehicle crashes that occurred in the wake of an emergency ambulance response was about four times higher than the number of ambulance-involved collisions,[11] so ambulances might be killing and injuring 300 civilians a year in the United States.

WHAT WE KNOW ABOUT THE RISKS TO EMS PERSONNEL

In this section, we will examine both fatal and nonfatal injury risks among EMS personnel.

FATAL INJURIES

A study from Australia found that there is no occupational group in that country with a fatality rate higher than that for paramedics; most of the fatalities were related to transportation events.[5] Figure 15.1 shows that in the United States, EMS personnel

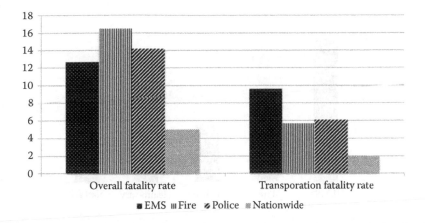

FIGURE 15.1 Fatality rates for EMS, firefighters, and police compared to nationwide average, overall and transportation related. US national data, 1992–1997.

have a fatality rate of 12.7 per 100,000 workers per year as compared to a rate of 14.2 for police and 16.5 for firefighters; if we look at just fatal transportation events, the rate is 9.6 for EMS workers compared to 6.1 for police and 5.7 for firefighters.[6]

Kahn et al. evaluated fatal ambulance crashes in the United States and found 592 nonfatal ambulance occupant injuries and 89 ambulance occupant fatalities during an 11-year period.[12]

Other causes of fatalities among EMS workers include assault and myocardial infarction. Each of those two causes results in an average of one to two paramedic fatalities per year in the United States. However, the infarction findings must be interpreted with caution because it is likely that only those events that occur on duty are included in occupational fatality databases; a paramedic who had a very stressful or physically demanding shift but did not arrest until arriving at home would likely not be included in those data.

NONFATAL INJURIES

Maguire and Smith found an average of 5400 injuries per year among EMS personnel in the United States.[9] Sprains and strains were the most common injury; the back was injured in almost half the cases. Falls and transportation events accounted for large percentages of the injuries; there were about 100 assault cases per year. Figure 15.2 shows data from a 2005 study that found that the rate of injuries among EMS workers was 34.6 cases per 100 full-time workers per year compared to rates of 18.6 for firefighters, 13.9 for police, and 6.9 for health services personnel and a national average of 5.8.[7]

In Australia, a 2014 study found that the risk of serious injury among paramedics is about seven times higher than the Australian national average.[5] That same study found that serious injuries that result in at least a week of lost work time result in an estimated 30,000 lost work days per year due to injury among the 15,000 ambulance personnel.

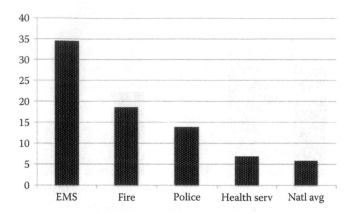

FIGURE 15.2   Nonfatal injuries. Number of injury cases per 100 full-time workers per year, 1998–2002.

Gershon et al. found 226 occupational injuries among a population of 197 full-time EMS workers in 1 year in Baltimore, Maryland; 15% of the cases resulted in over a week of lost work time.[13] The most common cause of the injury was "stretcher related." Reichard et al. examined injuries among US EMTs and paramedics; they found that stress, sprains, and strains were the most common diagnosis.[14] Studnek et al. found that the rate of reporting job-related illness or injury was associated with an urban versus a rural work environment, increasing call volume, and a history of prior problems with the back.[15] Heick et al. surveyed EMS personnel and found that 29% of respondents reported that they had been injured within the previous 12 months; of those reporting an injury, 64% reported multiple injuries.[16]

Back injuries tend to be the most common type of nonfatal injury reported among EMS personnel. A 2004 study of EMS personnel found that the rate of injuries related to lifting was 21 times higher than the national average and the rate of back injuries was 14 times higher than the national average.[17] Gentzler and Stader reported that "high risk was found for EMT patient care tasks that require reaching for overhead equipment or seated tasks that require horizontal bending and twisting."[18]

Transportation incidents lead to a large number of injuries among EMS personnel. Their risk of a transportation-related injury is five times higher than the US national average for all workers in the United States.[19] Becker et al. noted that "unrestrained ambulance occupants, occupants riding in the patient compartment and especially unrestrained occupants riding in the patient compartment were at substantially increased risk of injury and death when involved in a crash."[20] Saunders and Heye calculated crash injury rates for EMS personnel per 100,000 ambulance responses and found an overall collision injury rate of 22.2 injuries per 100,000 emergency responses.[21] Weiss et al. found that although there were more ambulance injuries in the urban environment, there was a greater severity of injury in the rural environments.[22]

Assaults and violence against EMS personnel result in over 100 serious nonfatal injuries among EMS workers each year in the United States[9]; for assaults that result in lost work days due to injury, the rate is 22 times higher for EMS personnel than the US national average.[17] Pozzi found that 90% of paramedics reported that they had been the victims of violence during their careers.[23] Corbett et al. reported that 25% of EMS personnel survey respondents reported being injured from an assault.[24]

When comparing to other emergency services personnel, Maguire et al. found that the injury rate among EMS personnel was 1.5 times higher than that among firefighters.[7] A 2014 study in Australia found that the risk of serious injury among paramedics is more than twice the rate for police officers.[5] Suyama et al. found that "EMS had higher rates of lost work and medical evaluations than both fire and police."[25] Reichard et al. examined ED treatment records and found that police officers and career firefighters had higher rates of ED visits than EMS personnel.[26]

Research has found that risks vary by demographic factors. In a 2004 report, Maguire et al. found that female EMS personnel had a higher injury rate than males, paramedics had a higher rate than EMTs, and the age group of EMS personnel with the highest injury rate was the 25- to 34-year-olds.[17]

Researchers have examined how fitness and health influence occupational risks. Studnek and Crawford found that EMTs who stated that they had good or fair fitness were three times more likely to report back problems than EMTs who stated that they had excellent fitness.[27] In another report, Studnek et al. found that "the variables most strongly associated with recent back pain and pain severity were prior back problems, self-reported health, and job satisfaction."[28]

IMPLICATIONS FOR THE WORKFORCE AND FOR OPERATIONAL READINESS AND COSTS

Ambulance services around the world are spending millions of dollars on injury prevention products and interventions but are not then evaluating the effects in a reliable enough way that the results can be published. So it is likely that many services are trying interventions that have been used in many other services without knowing if the interventions resulted in decreasing the risks or if they increased the injury rate.

Bulletproof vests are an example of an untested intervention. Anecdotal information indicates that some ambulance services in the United States have issued bulletproof vests to their EMS personnel. However, we have found no published studies documenting the results of using these devices. Without such studies, it is possible, and even likely, that the use of these vests actually increases the rate of violence against paramedics. Further, without research, there is no way to know of unintended side effects such as high rates of heat stress among paramedics using the vests.

POTENTIAL REMEDIES

This section will examine approaches to reducing occupational risks for EMS personnel. One of the challenges that we have when we consider occupational risk reduction initiatives for paramedics is that there is so little research on the topic. We will begin with the few papers that have been published.

Gamble et al. published the first EMS personnel injury prevention study.[29] In 1993, they had EMS personnel test subjects participate in a 10-week exercise study and evaluated the differences between the test subjects and a control group. They concluded that an exercise program "could prove cost-effective by increasing work capacity and decreasing absenteeism related to musculoskeletal injury."[29]

Three studies evaluated transportation-related risks. In 1997, Maguire and Porco examined interventions for reducing ambulance collisions; their multifaceted program of training and policy change resulted in a 50% decrease in ambulance collisions.[30] Levick and Swanson equipped test ambulances with an onboard computer that gave real-time auditory feedback alerting drivers to risky behavior such as driving without seat belts and speeding[31]; they found that the device led to a dramatic drop in violations and also resulted in significant vehicle maintenance cost savings. Myers et al. used a video recording device with a post-event follow-up review; they found that the program resulted in a decrease in negative driving "events."[32]

Two studies focused on needlestick injuries among paramedics. O'Connor et al. evaluated the effect of introducing self-capping IV catheters to a paramedic ambulance service; they found that the "incidence of contaminated needlesticks decreased from 169 to 0 per 100,000 IV attempts" per year following the intervention.[33] Peate evaluated the effect of introducing an automatically retracting lancet device and found that "needlestick injuries decreased from 16 per 954 EMS worker-years to 2 per 477 EMS worker-years."[34]

Studnek et al. found a significant decrease in the rate of back injuries among EMS personnel following the implementation of hydraulic stretchers.[35]

There are a number of initiatives that, although as yet unstudied, we expect could result in a decrease in occupational risks.

Driver Training and Evaluation

Haddon's matrix is a useful tool for determining the factors that may be related to ambulance crashes; it can help us focus on, for example, engineering changes such as "reducing diesel fumes, improving seatbelts, improving brakes and maintenance such as the frequency of checking inside rear tire pressure, and improving the safety of the patient compartment by eliminating protruding objects and sharp corners as well as adding padding and restraint systems for equipment."[36]

In the United States alone, there may be hundreds of different driver training programs ranging from a 1-hour lecture to over 40 hours of didactic and practical skill practice.[37] Although it seems intuitive that driver training programs result in safer drivers, we found no published research on the effect of any of those training programs. The lack of research leaves open the possibility that at least some of these ambulance driver training programs actually lead to higher crash rates than no training. We also have no research related to costs and benefits; this leaves open the possibility that a 1-hour class has as much benefit as a 40-hour class. Any ambulance driving training program should also "help the students develop the ability to do a risk–benefit analysis before turning on the emergency warning systems."[38]

Custalow and Gravitz evaluated 10 years of ambulance crash reports. Their suggestions include frequent training, ensuring that emergency vehicle drivers visually clear intersections before entering, make eye contact with other drivers, and use emergency response modes only when necessary.[39]

It seems both prudent and important that ambulance agencies screen potential employees' driving records and get annual updated motor vehicle citation reports on all personnel who may drive their emergency vehicles. Kahn et al. found that 41% of EMS drivers involved in collisions had poor driving records.[12]

FATIGUE

Anecdotal reports continue to suggest that some ambulance agencies are scheduling personnel for 24 and more straight hours of work. We now know that 21 hours of wakefulness produces impairment of the same magnitude as a 0.08% blood alcohol concentration (BAC)[40,41]; the legal limit for commercial drivers in the United States is 0.04% BAC.[42]

> In July 1997, Heather Brewster's car was rear-ended by Sookim Hong, a medical resident who had just finished a 36-hour hospital shift. Brewster suffered massive brain injuries and was in a coma for weeks. The accident left Brewster permanently disabled and the courts have declared her incompetent.[43]
>
> What makes this case particularly important to EMS agencies is that Heather's family sued the hospital, saying that it should have had at least as much responsibility as a bartender.

EMS agencies that schedule personnel for shifts in excess of 12 hours should reevaluate their scheduling patterns in light of the risks that long shifts pose to employees, to patients, and to the community.

CARDIOVASCULAR DISEASE RISK

Wolkow et al. described the result of a literature search that found cardiovascular disease (CVD) risk reduction interventions "which combined behavioural counselling, exercise and nutrition were more effective in improving cardiovascular health than nutrition, exercise or CVD risk factor assessment-based interventions alone. Further, CVD risk factor assessment in isolation proved to be an ineffective intervention type to reduce CVD risk."[44] EMS agencies and personnel should work together to reduce the risk of CVD among personnel.

STRESS AND SUICIDE

Safe Work Australia found that paramedics had the sixth highest rate of new stress claims.[45] In a 1988 study, Grigsby and McKnew tested paramedics and found "the highest mean burnout score yet reported for any group of health professionals."[46] Dutton et al. found the stress levels of paramedics to be significantly higher than the stress levels of firefighters.[47] Bentley et al. evaluated questionnaires returned by 34,340 EMS personnel and found that 7% were depressed; paramedics had a higher percentage of depression than EMTs.[48]

Hegg-Deloye et al. reported that "it seems clear that paramedics accumulate a set of risk factors, including acute and chronic stress, which may lead to development of cardiovascular diseases. Post-traumatic disorders, sleeping disorders and obesity are prevalent among [these] emergency workers. Moreover, their employers use no inquiry or control methods to monitor their health status and cardiorespiratory fitness."[49]

One report calculated that the rate of suicides among paramedics is 20 times higher than the rate among the general population.[50]

This is another issue that is best addressed through partnerships between ambulance agencies and personnel. It is possible that much of the problem is related to poor management and bullying in the workplace as well as by other work-related stress. Streb et al. concluded that enhancing resilience and "sense of coherence" were promising methods for reducing the risks of posttraumatic stress disorder.[51]

PERSONAL INTERVENTIONS

On a personal level, there are a number of practices that seem likely to result in decreasing an individual's risks of occupational injury and fatality.

In the ambulance, all occupants should be belted in at all times while the vehicle is in motion; it is likely that each time a provider removes his or her seat belt while in a moving ambulance, their risk of injury and the risk of injury to the patient and other compartment occupants likely increase.[52] For those rare times when it may be necessary for the medic in the back of the ambulance to unbelt in order to treat a patient, there should be a rule with the driver that the ambulance will slow down and perhaps not go over 25 miles/hour while the medic is unbelted. Emergency vehicle drivers should also remember that distractions contribute to the risks of collision.[53]

EMS personnel are often exposed to disease.[54,55] Vaccinations are critically important for all health workers and for emergency medical personnel especially. Get the yearly flu shot and all recommended vaccinations. See the Centers for Disease Control and Prevention website for a list of all recommended vaccines.[56]

Personal protective equipment can range from gloves to a full hazmat-like decontamination suit. EMS personnel should practice extra caution when dealing with unknown and potentially infectious patients. Gloves should typically be worn on all calls. Goggles, masks, gowns, and other protective equipment should be available and should be used whenever there may be a need.

Little is known about the circumstances of violence against EMS personnel.[57,58] Without specific data, the best we can say is that the risk of violence-related injury among EMS personnel is growing. Be aware of the risk, and if you feel that a particular call might be dangerous, call for a backup. Work with your partner(s) beforehand to develop coordinated plans for potentially dangerous scenarios. Periodically attend a self-defense class.

Stay fit, eat healthy, exercise, do not smoke, get sufficient sleep, and drink alcohol minimally, if at all. In addition to reducing risks of injury and helping toward a longer, happier life, Bentley et al. found that EMS personnel who described their health as very good had the lowest levels of anxiety.[48] And the healthier you are, the less likely you are to suffer a back injury.[15]

CONCLUSION

Risk reduction requires research, resources, and commitment from all levels of the EMS profession. The same methods that are used to ensure that medications are safe for our patients should be used to make sure that interventions are safe for our paramedics. EMS agencies should work with public health researchers to develop, implement, and test risk reduction strategies; the results of those analyses should be published so that EMS agencies around the world are not implementing, again and again, any ineffective interventions.

Databases must be developed for EMS research. The same systems that were developed to evaluate the differences in risks for office workers must be built upon and adapted for the unique needs of a profession that faces differences in risks on a moment-by-moment and call-by-call basis. The databases must be designed using standard constructs such as the Bureau of Labor Statistics' methods of classifying injuries by nature, body part, source, and event or exposure[59] and then building on this foundation to include EMS-specific coding such as those used by Maguire et al.[17]

Administrators concerned about the cost of implementing risk reduction programs can be reassured by research from other industries that shows that for every dollar invested in injury prevention, there is a three to six dollar saving in costs.[60]

Future research must also evaluate risks by demographic factors and by operational constructs. Important demographic factors include gender; research indicates that female paramedics have a higher risk of injuries (including assault and transportation-related injuries) than male paramedics, yet no one has studied the reasons for such differences. Operational constructs including scheduling patterns must be examined. It is likely that there are thousands of different scheduling patterns currently used by ambulance agencies around the world. It seems likely that some of them are associated with increased risks and some with decreased risks of occupational injury; we need to determine which ones are associated with lower risks.

The profession can become much safer than it is. Accomplishing that goal will require resources, commitment, and a team effort.

REFERENCES

1. Maguire BJ, Walz BJ. Current emergency medical services workforce issues in the United States. *Journal of Emergency Management.* 2004;2(3):17–26.
2. Harrawood D, Gunderson MR, Fravel S, Cartwright K, Ryan JL. Drowning prevention: A case study in EMS epidemiology. *Journal of Emergency Medical Services.* 1994;19(6):34.
3. O'Meara P, Tourle V, Stirling C, Walker J, Pedler D. Extending the paramedic role in rural Australia: A story of flexibility and innovation. *Rural and Remote Health.* 2012;12(2):1–13.
4. Comans TA, Currin ML, Quinn J, Tippett V, Rogers A, Haines TP. Problems with a great idea: Referral by prehospital emergency services to a community-based falls-prevention service. *Injury Prevention.* 2013;19(2):134–138.
5. Maguire BJ, O'Meara P, Brightwell R, O'Neill BJ, FitzGerald G. Occupational injury risk among Australian paramedics: An analysis of national data. *Medical Journal of Australia.* 2014;200(8):477–480.

6. Maguire BJ, Hunting KL, Smith GS, Levick NR. Occupational fatalities in emergency medical services: A hidden crisis. *Annals of Emergency Medicine.* 2002;40(6):625–632.
7. Maguire BJ, Hunting KL, Guidotti TL, Smith GS. Occupational injuries among emergency medical services personnel. *Prehospital Emergency Care.* 2005;9(4):405–411.
8. Zagaroi L, Taylor A. Ambulance driver fatigue a danger: Distractions pose risks to patients, EMTs, traffic. *Detroit News.* January 27, 2003. Available at: http://www.emergencydispatch.org/articles/driverfatigue.html. Accessed October 20, 2014.
9. Maguire BJ, Smith S. Injuries and fatalities among emergency medical technicians and paramedics in the United States. *Prehospital and Disaster Medicine.* 2013;28(4):1–7.
10. Maguire BJ. Ambulance safety in the U.S. *Journal of Emergency Management.* 2003;1(1):15–18.
11. Clawson JJ, Martin RL, Cady GA, Maio RF. The wake-effect—Emergency vehicle-related collisions. *Prehospital and Disaster Medicine.* 1997;12(4):41–44.
12. Kahn CA, Pirrallo RG, Kuhn EM. Characteristics of fatal ambulance crashes in the United States: An 11-year retrospective analysis. *Prehospital Emergency Care.* 2001;5(3):261–269.
13. Gershon RR, Vlahov D, Kelen G, Conrad B, Murphy L. Review of accidents/injuries among emergency medical services workers in Baltimore, Maryland. *Prehospital and Disaster Medicine.* 1995;10(1):14–18.
14. Reichard AA, Marsh SM, Moore PH. Fatal and nonfatal injuries among emergency medical technicians and paramedics. *Prehospital Emergency Care.* 2011;15(4):511–517.
15. Studnek JR, Ferketich A, Crawford JM. On the job illness and injury resulting in lost work time among a national cohort of emergency medical services professionals. *American Journal of Industrial Medicine.* 2007;50(12):921–931.
16. Heick R, Young T, Peek-Asa CL. Occupational injuries among emergency medical service providers in the United States. *Journal of Occupational and Environmental Medicine.* 2009;51(8):963–968.
17. Maguire BJ, Hunting KL, Guidotti TL, Smith GS. *The Epidemiology of Occupational Injuries and Illnesses among Emergency Medical Services Personnel.* ProQuest; 2004.
18. Gentzler M, Stader S. Posture stress on firefighters and emergency medical technicians (EMTs) associated with repetitive reaching, bending, lifting, and pulling tasks. *Work.* 2010;37(3):227–239.
19. Maguire BJ. Transportation-related injuries and fatalities among emergency medical technicians and paramedics. *Prehospital and Disaster Medicine.* 2011;26(5):346–352.
20. Becker LR, Zaloshnja E, Levick N, Li G, Miller TR. Relative risk of injury and death in ambulances and other emergency vehicles. *Accident Analysis & Prevention.* 2003;35(6):941–948.
21. Saunders CE, Heye CJ. Ambulance collisions in an urban environment. *Prehospital and Disaster Medicine.* 1994;9(2):118–125.
22. Weiss SJ, Ellis R, Ernst AA, Land RF, Garza A. A comparison of rural and urban ambulance crashes. *The American Journal of Emergency Medicine.* 2001;19(1):52–56.
23. Pozzi C. Exposure of prehospital providers to violence and abuse. *Journal of Emergency Nursing.* 1998;24(4):320–323.
24. Corbett SW, Grange JT, Thomas TL. Exposure of prehospital care providers to violence. *Prehospital Emergency Care.* 1998;2(2):127–131.
25. Suyama J, Rittenberger JC, Patterson PD, Hostler D. Comparison of public safety provider injury rates. *Prehospital Emergency Care.* 2009;13(4):451–455.
26. Reichard AA, Jackson LL. Occupational injuries among emergency responders. *American Journal of Industrial Medicine.* 2010;53(1):1–11.
27. Studnek JR, Crawford JM. Factors associated with back problems among emergency medical technicians. *American Journal of Industrial Medicine.* 2007;50(6):464–469.

28. Studnek JR, Crawford JM, Wilkins J, Pennell ML. Back problems among emergency medical services professionals: The LEADS health and wellness follow-up study. *American Journal of Industrial Medicine.* 2010;53(1):12–22.

29. Gamble RP, Boreham CA, Stevens AB. Effects of a 10-week exercise intervention programme on exercise and work capacities in Belfast's ambulance-men. *Occupational Medicine (Oxford, England).* 1993;43(2):85–89.

30. Maguire BJ, Porco FV. EMS and vehicle safety. *Emergency Medical Services.* 1997;26(11):39–43.

31. Levick NR, Swanson J. An optimal solution for enhancing ambulance safety: Implementing a driver performance feedback and monitoring device in ground emergency medical service vehicles. Paper presented at: Annual Proceedings/Association for the Advancement of Automotive Medicine 2005.

32. Myers LA, Russi CS, Will MD, Hankins DG. Effect of an onboard event recorder and a formal review process on ambulance driving behaviour. *Emergency Medicine Journal.* 2012;29(2):133–135.

33. O'Connor RE, Krall SP, Megargel RE, Tan LE, Bouzoukis JK. Reducing the rate of paramedic needlesticks in emergency medical services: The role of self-capping intravenous catheters. *Academic Emergency Medicine.* 1996;3(7):668–674.

34. Peate W. Preventing needlesticks in emergency medical system workers. *Journal of Occupational and Environmental Medicine.* 2001;43(6):554–557.

35. Studnek JR, Crawford JM, Fernandez AR. Evaluation of occupational injuries in an urban emergency medical services system before and after implementation of electrically powered stretchers. *Applied Ergonomics.* 2012;43(1):198–202.

36. Maguire BJ. Preventing ambulance collision injuries among EMS providers: Part 2. *EMS Manager and Supervisor.* 2003;5(3):4–7.

37. Maguire BJ, Kahn CA. Ambulance safety and crashes. In: Cone DC, ed. *Emergency Medical Services: Clinical Practice & Systems Oversight:* NAEMSP; 2009.

38. Maguire BJ. *Characterizing ambulance driver training in EMS systems.* US DOT Docket ID Number NHTSA-2014-012. 2015. Available at: https://www.regulations.gov/document?D=NHTSA-2014-0127-0002. Accessed February 17, 2015.

39. Custalow CB, Gravitz CS. Emergency medical vehicle collisions and potential for preventive intervention. *Prehospital Emergency Care.* 2004;8(2):175–184.

40. Arnedt JT, Wilde GJ, Munt PW, MacLean AW. How do prolonged wakefulness and alcohol compare in the decrements they produce on a simulated driving task? *Accident Analysis & Prevention.* 2001;33(3):337–344.

41. Dawson D, Reid K. Fatigue, alcohol and performance impairment. *Nature.* 1997; 388(6639):235.

42. US National Highway Traffic Safety Administration. *Commercial drivers license.* Available at: http://www.nhtsa.gov/people/injury/enforce/cvm/cmv_license.html. Accessed October 19, 2014.

43. Gotbaum R. Safety of medical residents' long hours questioned. *NPR Health & Science.* February 28, 2005. Available at: http://www.npr.org/templates/story/story.php?storyId=4512366. Accessed August 27, 2013.

44. Wolkow A, Netto K, Aisbett B. The effectiveness of health interventions in cardiovascular risk reduction among emergency service personnel. *International Archives of Occupational and Environmental Health.* 2013;86(3):245–260.

45. Safe Work Australia. *Compendium of workers' compensation statistics Australia 2004–05.* 2007. Available at: http://www.safeworkaustralia.gov.au/sites/SWA/about/Publications/Documents/423/Compendium_Workers_Compensation_Statistics_Australia_2004_05.pdf. Accessed April 22, 2012.

46. Grigsby DW, McKnew MA. Work-stress burnout among paramedics. *Psychological Reports.* 1988;63(1):55–64.

47. Dutton LM, Smolensky MH, Leach CS, Lorimor R, Hsi BP. Stress levels of ambulance paramedics and fire fighters. *Journal of Occupational and Environmental Medicine.* 1978;20(2):111–115.

48. Bentley MA, Crawford JM, Wilkins J, Fernandez AR, Studnek JR. An assessment of depression, anxiety, and stress among nationally certified EMS professionals. *Prehospital Emergency Care.* 2013;17(3):330–338.

49. Hegg-Deloye S, Brassard P, Jauvin N, Prairie J, Larouche D, Poirier P, Tremblay A, Corbeil P. Current state of knowledge of post-traumatic stress, sleeping problems, obesity and cardiovascular disease in paramedics. *Emergency Medicine Journal.* 2013;31(3):242–247.

50. Bucci N. Alarm at suicide for paramedics. *The Age.* May 2, 2012. Available at: http://www.theage.com.au/victoria/alarm-at-suicide-for-paramedics-20120501-1xx8m.html. Accessed May 11, 2012.

51. Streb M, Häller P, Michael T. PTSD in paramedics: Resilience and sense of coherence. *Behavioural and Cognitive Psychotherapy.* 2014;42(4):452–463.

52. Maguire BJ. Another Look at the emergency medical services safety report. *Annals of Emergency Medicine.* 2003;42(1):159.

53. Maguire BJ. Ambulance Safety. In: Cone DC, ed. *NAEMSP. Emergency Medical Services: Clinical Practice and Systems Oversight*: Wiley; 2014.

54. Makiela S, Taylor-Robinson AW, Weber A, Maguire BJ. A preliminary assessment of contamination of emergency service helicopters with MRSA and multi-resistant *Staphylococcus aureus. Emergency Medicine: Open Access.* 2016;6(1):304. Available at: http://www.omicsgroup.org/journals/a-preliminary-assessment-of-contamination-of-emergency-service-helicopters-with-mrsa-and-multiresistant-staphylococcus-aureus-2165-7548-1000304.php?aid=66772.

55. Amiry AA, Bissell RA, Maguire BJ, Alves DW. Methicillin-resistant *Staphylococcus aureus* nasal colonization prevalence among emergency medical services personnel. *Prehospital and Disaster Medicine.* 2013;28(4):1–5.

56. US Centers for Disease Control and Prevention. Recommended vaccines for healthcare workers. Available at: http://www.cdc.gov/vaccines/adults/rec-vac/hcw.html. Accessed October 19, 2014.

57. Maguire BJ, O'Meara P, O'Neill BJ. Violence against paramedics: Developing the tools to end the epidemic. *Response.* 2016;43(1):24.

58. Maguire BJ, O'Meara P, O'Neill BJ, Brightwell R. Violence against emergency medical services personnel: A systematic review of the literature. *Prehospital and Emergency Care* (in revision). 2016.

59. US Bureau of Labor Statistics. *Occupational Injury and Illness Classification Manual 2012.* Available at: http://www.bls.gov/iif/oiics_manual_2010.pdf. Accessed October 19, 2014.

60. Oster KV. Return on investment for employee safety programs. *Houston Chronicle.* Available at: http://smallbusiness.chron.com/return-investment-employee-safety-programs-47715.html. Accessed October 19, 2014.

Index

Page numbers followed by f and t indicate figures and tables, respectively.

Milton Keynes UK
Ingram Content Group UK Ltd.
UKHW040107071024
449327UK00019B/872